MY INCARCERATED LIFE

The Intimate Details of a Prison Wife

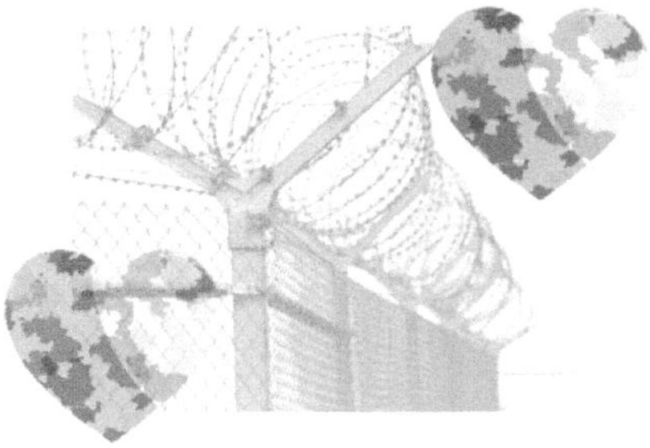

Qiana Riley-Jones

****

Qiana Compositions, LLC
PO Box 7018, Goodyear, AZ 85338

This is a work of nonfiction. A memoir.
Some names and identifying details of people
described in this book have been altered to protect their privacy

 First paperback edition June 2021

Book design by Qiana Riley-Jones

Book club inquiries or speaking engagements can be submitted via email.
Our books may be purchased in bulk for promotional, educational,
or business use.

Please contact bookseller or the self-publisher Qiana Riley-Jones
of Qiana Compositions, LLC
by email at qiana.compositions@gmail.com.

ISBN 978-1-7371678-0-8 (paperback)

Manufactured in the United States of America
Published by Qiana Riley-Jones

www.qianacompositions.com instagram.com/qianacompositions
twitter.com/qianacompositions facebook/qianacompositions

****

TABLE OF CONTENTS

PROLOGUE

The Verdict's In…

The shrill sound of a phone echoed through the kitchen. Qiana looked at the caller ID and recognized the number as the same 925-area code that she had come to see more frequently. She answered and a disembodied voice on the other end said, "You have a collect call from "Boo."

"I accept the charges," Qiana immediately replied.

"Hey stranger, what's up with you?" Vandrick's voice on the other end sounded unsettled, but confident.

"Nothing much. I was in the kitchen," Qiana told him.

He gave a slight laugh. "I bet you are. I still can't believe you're pregnant."

"Yeah, I can't either," she responded quickly.

"Umm-hmm. It don't matter. I'm still your first baby's daddy." He paused for a moment then said, "Well, the trial is almost over. Not too much more left."

"So how do you think this is going to go?" Qiana inquired.

"I'll be fine. They don't have any evidence. But even if they try to put this on me, my folks say they'll stand up and take these charges because they know I didn't do this."

Qiana listened attentively. She knew nothing about what was going on, so she had no choice but to take his word for it, even though it didn't sound realistic. "Do you want us to come to court?" she asked.

"Nah, do what y'all got to do. Everything is going to be fine. Where's my son at?"

"In his room."

"Let me holler at him."

"Hold on," she told him and called out for their son to come to the phone.

He walked into the kitchen and took the phone to his bedroom where he was playing his video games and started talking to his dad. He returned after only a few minutes of conversation and handed her the phone saying, "Here mom, he wants to talk back to you."

She took the phone and they made small talk, catching up a bit before he ended his call. As the call ended, she told him, "I'll be praying for you."

"Thank you," he replied in a subdued voice.

It felt like it had been at least two weeks since she last heard from Vandrick. Yeah, she had never called him by his street name, "Big Nose Boo" or "Boo," a name of affection his mother and big brother, Ron, had given him. Qiana always felt that name was reserved for the people that knew him as such. She only knew it as a street name and since she wasn't a part of the streets there was no need for her to become familiar with it. Vandrick worked best for her and he had no issues with her using it. Maybe it hadn't been two weeks, but it definitely felt like a long time had passed as she waited to find out from him what was going on.

It wasn't long before her phone rang again and it was that 925-area code. She was a bit puzzled because if he was found not guilty wouldn't he be calling from his cell phone number? But she accepted the charges and decided she would let him talk to her versus assuming anything. The call was definitely awkward. Instead of the more jubilant tone of someone who had won their case, the phone call she was greeted with had an eerily hushed and almost shocked quality.

"Hey, how are you?" Qiana cautiously asked.

"I'm doing okay," Vandrick answered.

"How are you really doing?" She asked again.

"Qiana, they said I'm guilty," he answered in a stunned voice.

"Wait…what?" It was taking her some time to process what he'd just told her.

"Guilty," he repeated. Qiana couldn't believe it. It was still not registering with her. Vandrick continued, "How did this shit happen to me?"

"Vandrick," Qiana's voice changed from confused to concerned, "How are you holding up? Do you need anything?"

"I'm not good," he replied. "I've kept mostly to myself. I am not processing anything. I just gotta wait 'til sentencing."

"What is that looking like?" Qiana asked, almost not wanting to know.

"For this… 40 years to life."

Now she was stunned. She didn't want to ask too many questions but her mind was racing with thoughts. This was not the time to drill him with questions. His life had to surely be flashing before his eyes. Qiana started to check her reactions and questions, remembering to never ask questions or talk about things that you didn't want recorded. So, she couldn't ask for any details. Vandrick sensed her thoughts because of what he said next.

"The case was built against me…" That phrase and the next one would remain with her for many years to come. "… and the public defender did nothing to prove my innocence."

Qiana remained silent for a moment because of all the thoughts still circulating in her brain, trying to determine what he was thinking. The only thing she could ask that made any sense to her at the time was, "Do you need me to come to court for your sentencing?"

He replied again, "No. I want you to stay far away from this stuff. It's ugly!"

"How is your wife handling all of this?" She'd almost forgotten to ask.

"She lost it in the courtroom. She is not doing okay at all," Vandrick said.

"I'm here if you need anything," Qiana assured him.

"Yeah, I know. I'll call you back once I know more," he replied.

"Take care, Vandrick," she told him, really meaning it.

"You too," he said.

After that call, Qiana didn't know what to do. Does she talk to their six-year-old son and invite him into her world of confusion where, at his age, he would have a lot of questions that she had no answers to? Or, does she wait and try to guide him through this maze of uncertainty as she learns more? Qiana is often left to navigate through the most tangled of obstacles and this would be no different. But he was a child so she would have to guard his young mind as much as she possibly could; especially as the thoughts circulated about that first time, she ever had to take him to visit his dad in the county jail. She had been there before, visiting an ex-boyfriend in the past and learned the process from that experience. However, she had never thought about it from the view of a child entering this callous environment and the effects associated with visiting a parent in jail. That part she should have given more consideration to, but Qiana had yet to even take in what had now become their dismal reality.

CHAPTER ONE
Just Kickin' It

12 Years Earlier…

I was 14 years old when I met Vandrick and he was 16. I lived on the east side of Oakland, California. There, the environment was quickly deteriorating due to the surge in drugs and violence. To be honest, programs such as D.A.R.E were not having the anticipated positive impact in the area where I was being raised. And I admit Vandrick was a part of the contribution to the rapid decline of my neighborhood. He was one of the street dealers. I overlooked the fact that his choice of lifestyle wasn't like mine. I didn't know what it was, but there was something about him that seemed a bit different than other drug dealers I had encountered. It helped that he had a southern accent when he'd asked for my name and phone number as I was making my way home from school.

I asked him his name and he confidently responded, "Boo."

Then I looked at him with a side eye and a slight grin, asking, "So, you mean to tell me your mother named you, Boo?"

He smiled and replied, “No.”

“Then what is your real name?”

“Vandrick,” he replied almost bashfully.

I guess he didn’t find that name as street savvy as the nickname, Boo. It was also apparent many girls had never challenged him to reveal that bit of information, but he needed to know I was not your average girl. This scene would take place a few times because he lost my phone number two times after our first encounter. It was kind of cute, in my adolescent mind, to see him admit that yet again he’d lost my number. I’d grin as I saw the look of shame on his face when he’d tell me he’d lost the number, but intrigued at the same time that he never forgot my name. Qiana wasn’t as common as most names like Kellie or Keisha.

So, by the third time, I took the initiative and requested he just give me his phone number and I would call him. He did and later that evening after I finished my homework, I built up the needed confidence to dial his number. It rang a few times and then a young male voice answered. I thought it might be him but didn’t want to assume it was because I could hear a few voices in the background.

“Hello, may I speak to Vandrick?” I asked.

The male voice replied with uncertainty, "Who?"

I thought at that moment maybe I'd dialed the wrong number. So, I hesitantly repeated, "Van…drick?"

"Oh, hold on. Boo, telephone!" the boy yelled out.

"Hello," Vandrick answered with some apprehension in his voice.

"Hi, Vandrick. How are you?" I asked.

It took Vandrick a minute to figure out who would call him by his birth name until he quickly realized who it was. "Oh! Hey, Qiana. I'm okay and how are you?"

We spent some time on the phone, conversing and getting to know each other a bit better. He seemed as nice as the tone I had heard in his voice when we first met, so my assumption was confirmed. He was definitely from the South. He was originally from Louisiana and hadn't been in California very long. I was intrigued by him, but wouldn't press any issues. He seemed like a cool person who I could enjoy having as a friend. I said friend because he let me know in our conversations about his girlfriend. He referred to her only as Tiff'.

After my initial call to his house, I didn't make too many more calls. It was easier to talk to him on the block as I headed home. Now that he had my number it was no longer something he could say he'd lost. I would often sit on my front porch and hang out with my other friends in the neighborhood. They all knew Vandrick because many of them went to the same school, that is, whenever he decided to go. However, my mother didn't allow me or my older sister to go to the schools in the surrounding area, so we commuted to the more suburban areas to receive our education. Unfortunately, this was what some parents felt was absolutely necessary to try to avoid the potential pitfalls for their children due to the negative changes in the environment within their neighborhood. Both my sister and I spent long trips on buses to get back and forth to school. Once I got back to the neighborhood and did my homework, if it wasn't too late in the evening, I could hang out but only for a little while. However, I usually had time to hang out with my friends for longer periods of time on the weekends.

I can't really recall how Vandrick and I determined the dynamics of our friendship, but it didn't take long for us to become what was commonly referred to as "friends with benefits." He was not my boyfriend and I definitely was not his girlfriend. We were certain about that, but there was a sexual chemistry between us that neither of us fought very hard to disregard. During this time, I found out more about him and he found out more about me from our random "hooking ups." We never discussed with others what we had going on, at least I knew for sure I hadn't. It wasn't a secret, but I wasn't one to put everyone into my business either. I always had a way about me that said, "What I have going on is on a need-to-know basis. If you don't need to know, I'm not going to freely share it." Vandrick and I had this compatible synergy between us which worked out well. We had many similarities which contributed to our physical connection, even if it was only for a couple of hours in the still of the night.

Vandrick's drug dealing and reputation was increasing. I noticed he wasn't just hanging out and racing against the other guys in the area to make the first sale anymore. He had made enough money to buy himself a car. It was a blue Caprice Classic with all the bells and whistles, including the amps and subwoofers that would rattle the ground as he came cruising down the block. I knew he would soon drive down my one-way street because I'd heard his favorite song, "Ain't No Fun" by Snoop Dogg. Whenever I heard that song in the distance, I knew it was him but didn't know if he was coming by to see me, or if he was in the area to take care of some business.

I never questioned him. I knew how to mind my own business, plus he always followed our "call before coming by" policy. This was our reality for years, even when I was no longer living in the area. We'd lose contact, misplace numbers and not speak for months at a time, but it never stopped us from having that random hookup; thanks mainly to my sister always going back to visit the old neighborhood because my mom still owned our house on the block. We never lost connection with the area. After one of her many visits back to the neighborhood, it never failed, my sister would approach me with the same question every time.

"Hey, I was in the neighborhood and guess who I saw?"

It had become so common I anticipated her asking me this so I just rolled my eyes and with a grin replied, "Who, Vandrick?"

And as expected, she replied, "Yes, and just in case I forget to tell you, I gave him your number. He told me to tell you to call him."

I would act on the request when I had time. The memories of the many nights we had, locked in the grip of each other's bodies, kept the expectation of another great experience in my constant thoughts. There was nothing like knowing that the sound of the bass coming from inside his car's trunk a few blocks away was soon to be Vandrick, lowering the volume as he pulled up in front of the house. He would come up the stairs to enter into our own space and time; had just enough mystery intertwined with the lust from the magnetic chemistry that existed just between the two of us. He would casually walk around—his confidence always preceded his entry—getting comfortable as he ended his long day of hustling, gambling, or God knows whatever else. I prepared a meal for him, wearing the most accessible attire I had, knowing we were not in each other's presence to shoot the breeze but to release every ounce of forbidden pleasure upon each other. Once completion was reached, we would end up wrapped in each other arms, falling sound asleep until the sun slightly crested the horizon and gradually brightened the sky. We also knew this was where I would wake up and remind him that it was time for him to go and I could start my day.

Years went by with these random hookups—four years to be exact. We never questioned our relationships with others or things like that. What he did with others outside of our friendship was his business and the same rule applied for me. We established this rule very early and allowed it to remain without a need to address it again. This worked well for us because there was no need for unnecessary questioning when all that was taking place was hooking up.

I had been away from Oakland for some time and at 18-years old I asked my mom if I could become her tenant and move back to the old neighborhood house in Oakland. She was in agreement with a few stipulations. I was willing to accept the terms and I made sure to give her my rent money up front because I had a decent paying job downtown. I didn't have a lot of money to hit the world running, but I wanted to come as close as I could to providing for myself and have some freedom. I felt it was time because in addition to Vandrick's hookups, so many things were happening for me as I transitioned from a teenager to a young adult, including declining a few offers to go away to college. Love and life seemed far more appealing than packing suitcases and leaving behind what appeared to be my entry into adulthood.

I was now in a relationship with a guy I had been involved with for over a year, my high school sweetheart. It was my first really serious relationship. We were preparing to get married, start a family and everything. I had placed school on the back-burner for myself as I waited for him to complete his senior year. But things would not work out according to either of our plans as we suffered some challenging circumstances along the path that none of us were equipped to handle. It all happened so fast, one devastating blow after another, destroying the fibers of the yearlong love and friendship we had invested in. The saying was true that "When you think all is perfect, you will be tested."

We suffered through a major argument about him getting ready to make the biggest mistake of all time. He constantly sought ways to make quick money and being around the drug dealers in the neighborhood— as he had come to stay with me on the block, commonly referred to as "The One-Way" —his attempt to go down that path instead of following our original plan of him completing high school and enlisting in the military, led to me calling off the wedding. I experienced an abrupt miscarriage and eventually accepted the offer of sex to console myself, ending up in bed with my reliable friend-with-benefits. It was a whirlwind of turbulent events that seemed to last an eternity for me but in reality, happened all within a matter of a couple of weeks. I was trying to process through all of my anger and the frustration of calling off a wedding that was three days away from the ceremony.

Then I received an unexpected call from Vandrick on his way to a Halloween party asking, "What are you up to?" At the time that was nothing more than me lounging around the house, watching TV. Instead, I ended up with the two of us in my bed, relieving some of my built-up tension. A few weeks later this spontaneous hookup resulted in me being in the hospital, confused and pregnant by my friend-with-benefits, Vandrick. Completely unprepared, this was a conversation I never wanted to relive again in my life, but it was the most mature thing I could do. It taught me that no matter what, I was strong enough to weather through a tsunami of emotions and I held myself accountable for my actions and any hurt I might have caused along the way.

I called both of them over to the house and made sure all the details and time lines of events were clear. It was a challenging time and a conversation that I was unprepared to deliver. However, even beyond their individual frustrations, both guys respected the guts I had and my integrity. I didn't cheat on my ex-fiancé, but the timing of everything was too close for comfort. My anger and lack of judgment played a key role in my bad decision-making and I owned up to it, especially since shortly after the night of Halloween sex and the uncomfortable confession, my ex-fiancé and I decided to try to salvage our relationship to the best of our ability. Not to the point of marriage, but back to re-evaluating the option of a relationship. I was guilty of being a reckless teenager with misguided realities of being an adult, having fun but not thinking about the consequences and repercussions that were to come from my actions; but I was willing to take responsibility. My intent was never to hurt anyone, but many lives were affected and turned upside down after one night of pleasure.

What also floated to the surface during this time was that Vandrick was already a father. I already knew this because a family member of his told me one day in class before I graduated high school when she realized I knew her cousin. However, on the night I met with him and my ex-fiancé, he let me know he had another child, a son, on the way in just a few more months. He said he was not ready to be a father of three at 20-years old. It was difficult to process as I heard what he said, but it bore no weight on my decision to continue my pregnancy. Everything at this point was all spontaneous, explosive and confusing. I was not asking him to change his situation based on what was now taking place with me. I had enough to deal with on my own table and I knew I did not want to think about him, too. Whatever he needed to deal with in his outside world I trusted he would figure it out on his own. I just wanted him to be aware that I was pregnant with his child.

I had witnessed people in the past keep longtime secrets regarding who they were having sex with. They would complete entire pregnancies without ever admitting the facts; and children were born believing they were fathered by one man when their father was someone totally different. I was certain I did not want this to be the case for me. Vandrick deserved to know the truth and our baby was not going to be born living a lie. There was enough confusion and me intentionally adding to it was not about to take place. I was not at all suggesting he leave whatever he had going on, or him becoming my boyfriend all of a sudden. So, after that grueling night, I spent three trimesters with mixed feelings, emotions and the complexity of being pregnant throughout that time. I had to readjust my entire life's plan because soon enough our son was born.

CHAPTER TWO

Jet Ski and Delivery

The day our son was born, Vandrick was on an annual camping and boating trip that many people traveled to in the summer at Lake Berryessa. He was sure that I would not have the baby on my actual due date. Boy! Was he wrong! I cooked his favorite meal of macaroni and cheese and fried chicken that he craved before he left. He came by and ate; we chatted a bit and he headed out on his long weekend of fun in the sun. When I was in active labor, I asked my sister to call my high school sweetheart. We had recently gotten into another argument after my baby shower a month prior, but I wanted his support since we had spent the entire eight months together during my pregnancy, trying to work through things. I was looking forward to seeing him more than anyone else because I also realized it was getting close to the date for him to leave for boot camp. However, my sister decided to call Vandrick instead. What I remembered through the pain I was tolerating— and the blur of strong medication the nurse injected into my veins against my better judgment—was my sister saying, “He’s on a jet ski in the middle of the lake and won’t be able to get to the hospital anytime soon.”

I don’t know if I cared or not, I just knew I was in lots of pain and I needed to focus because the full 22 hours of labor was far more real than his presence. The medication was having an intense effect on me and I couldn’t keep my eyes open long enough and my words clear enough to yell at her, “I never asked you to call him! I know where he’s at!” There were plenty of people in the hospital to coach me through labor and delivery and I didn’t need to be focused on his absence, especially while he was out having fun on the lake. The pain was so extreme that my mother, during my delivery, had the biggest laugh as I told the Midwife who was doing all she could to help in the process of delivering the baby who I now believed had the biggest head imaginable, “Please, let’s stop and I will come back in two weeks!” Not to mention the fact that I was born two weeks past my due date and my mother said her pain was worse than mine. She had back labor and attempted to relieve the pain by doing a full handstand off the side of her sofa before I was born.

Finally, our son was born. He was a healthy and happy bundle of joy brought forth through the turmoil of his conception and the pain of his delivery. There was finally calm after the storm. Everything subsided as I looked into his squinted little eyes. I promised him I would do everything in my power to raise him to the best of my ability. I was a child raised in a single-parent household, so I knew that it was possible to take on this responsibility. I placed no obligations on Vandrick, and I also knew I would never be the type of mother to deny him access to his child just because what we had between us was undefined and suffered through some glitches. Our son had no burden to bear for the poor adult decisions we made and the actions we participated in. Vandrick had every right to participate in his son's life to whatever level he could willingly maintain.

Vandrick returned from his out-of-town camping trip and he and another one of his friends dropped by the house. I was in my room, giving our son a sponge bath and preparing him for bed. He had been born bright and early on a Monday morning and it was now Tuesday night and I was back home, learning my routines as a new mommy. My cousin, who was pregnant with her first child at the time, was lying on my bed half asleep, watching me. After she let Vandrick and his friend into the house they found their way to where I was bathing the baby. Vandrick and his friend stood in the doorway of my bedroom behind me with the baby facing them so I couldn't see their expressions. I looked over my shoulder and greeted each of them, and they both responded, "Hey," almost in unison. But after that there was absolute silence. I noticed that his friend had left the doorway and gone into the living room and sat on the couch in the dark. He probably dozed off while Vandrick was still standing there, silent enough to hear a pin drop, staring at his son I was holding in my hands and caressing gently with his bath towel across his body. He lay there, silently enjoying the warmth of the room and his first baby bath-time massages of which there would be many.

"Are you okay?" I asked Vandrick.

He quietly replied, "Yes."

"Would you like to hold him?"

He seemed a bit puzzled at that question, as if trying to determine whether he would interrupt what I was doing or let me finish. He declined, "No, finish what you are doing. I'll come back tomorrow."

They were clearly tired from their weekend. Vandrick was complaining about his back hurting and he needed to drop his friend off as well. He'd just wanted to come straight into town and see the baby, his now second son. An undeniable fact at this point as anyone with eyes could see was that his son was the spitting image of him and looked nothing like me. As was commonly said, "Girl, you had nothing to do with having this child; you only carried him."

I finished putting our son in his pajamas and swaddled him backup to prepare for his nighttime feeding and let him drift back off to sleep. One thing was for certain, Vandrick never stopped dropping by to visit. And, despite the new addition of our son into an unpredictable relationship, we always remained cordial with one another. Our sexual encounters after our son was born had tapered off a bit. I don't know which one of us contributed to this change, but as he put it, "I now have baby daddy privileges." So, the sex wasn't completely out of the question, but it was less frequent than before. I never took the things he would say to me seriously, but in the back of my mind I knew it would be our reality. No matter how often he would say them, I would adamantly deny the things he would say to me.

Vandrick was still active in the streets, gaining an even stronger presence and moving up into the ranks. He treated the street life as a real business. As I had said, I never asked questions because I never wanted to know. What I was exposed to growing up in the neighborhood was enough for me to narrow down that he lived a fast and dangerous life. The less I knew about the specifics the better off I was. That was not the lifestyle I lived so what he did outside of my sight and presence I did not question. I raised our son, worked a regular job and provided our son a lifestyle that did not mirror that of his father's. He would follow in my upbringing. I sent him to affluent schools no matter where I lived, even if that meant paying for private school education.

Periodically, Vandrick would ask, "Can I come by and pick up my son?" when he was planning an outing with him and his other children. I never opposed his request. Our son had siblings and I wanted him to know that. I always had the mentality that I didn't want our son to grow up only knowing my family and one day run across a female family member he might be attracted to and unknowingly date a relative. The siblings' time together and their knowing each other was important to me. It wasn't unusual for there to be gaps in time where I wouldn't see or hear from Vandrick. However, some way or somehow, he always knew how to get back in contact and he always relied on my mom's house in the neighborhood to find me.

There were a few times he had gone to jail for various things, but from what I recalled he never stayed inside too long. He always made bail and adhered to court dates; did a short sentence in the county jail and he'd get out. He even asked me once or twice to sign for him to be released on bail since I was working. That time was a bit nerve-wracking for me because I had to put my trust in him that he wouldn't skip bail. He didn't! He kept his word. I appreciated that. Otherwise, the burden of his failures would have ultimately fallen on me.

Our son had his commencement ceremony from a private preschool in 2002. We all showed our support: my mom, step-dad, Vandrick, his friend who was always with him and my boyfriend who our son started to enjoy being around. It was a bit awkward for my boyfriend and my child's father to mingle in those close quarters, but it was only for a short period of time and it would be over soon. I never brought my son around anyone I wasn't serious about so for my boyfriend to be there and our son to even know him meant that he was not just some guy that meant little or nothing to me. It was a cute ceremony right in the play area of the preschool. There were little people in white caps and gowns, performing a short skit with some songs, a few memorized sayings with a presentation of miniature diplomas. We all gathered inside the school and had refreshments and snapped a few pictures of our graduates and their friends. We even had a chance to take a photo of the three of us: our son, me and his father. We didn't have many of those, but this was a rare occasion so we stood for the moment.

I'm not sure what happened outside, but I could see by the look on Vandrick's face that he wasn't the happiest of campers. I was certain I would find out later in the day what was going on, if I needed to know. I hadn't heard there was any fight that had broken out, so I am okay with not knowing for the time being. This moment is about our son and his achievements, not adults and their personality clashes. I later found out from my mother that there was a seat assignment discrepancy between Vandrick and her, but just as I thought, it wasn't a major emergency. Did anyone even notice I was his mother and I stood during the entire ceremony? So, if they were feuding over a chair then they would likely figure it out. The version I heard from Vandrick was that he and my boyfriend had an issue. My boyfriend, who was spending time with our son, seemed to feel his dad didn't need to attend. I don't know what all was transpiring as I was taking in the ceremony, but what I wasn't going to do was ruin our son's day over adults behaving badly.

I eventually moved from the old neighborhood and was living in an apartment complex my boyfriend owned. I was a regular tenant just as everyone else. Yes, he owned it, but I made sure to pay my rent regardless of what relationship we had. If our relationship took a turn for the worst, he would not be able to put me out based on whatever our quarrel was. He would have to go through legal action as any landlord and tenant required. It was business over pleasure in my book. During this time, I didn't see Vandrick often and the friend-with-benefits portion of our relationship had come to a halt. We communicated on behalf of our son, but that was about it. At this point he had three more children. I recalled him calling one day and he and I jokingly referred to how many children he had at the time. He replied, "You know I am a man from the South and we have big families. My plan is 20. I will get to 10 and then go back to each woman and have another with her, so you need to get ready."

This is why I never take him seriously. He was with the jokes and I know for sure that wasn't happening. And if he was for real, I could *guarantee* I was not showing up for my second time around. He really called to invite our son to his youngest daughter's first birthday party. He wanted to get all his kids together and spend time with them. Of course, I had no issues with it. He asked me to drop our son off at his friend's—who was always with him—father's house so the kids could put on the outfits that he had bought for all of them. He wanted them to dress alike. It was common for him to make sure his babies were all dressed alike when he had them with him. I told him I would get our son there on the day and time he requested.

The day of the party arrived our son had been out earlier with my boyfriend, getting their hair cut, and I was waiting for them to return. It was getting later in the day and close to the time for me to drop off my son. Vandrick called to make sure I was still bringing him to the house. I told him I was waiting for my son and boyfriend to return and once they got back, I would bring him. He wasn't happy about it, but what could I do? I didn't have our son with me so I had to wait for him to return. They eventually showed up and I made sure he was cleaned up. But because of their lateness, now I needed to take him straight to the birthday party a couple of cities over at the park.

When I got there, Vandrick was less than receptive to the fact that our son was not dressed in the same outfit as the other kids. Our tardiness was the cause of this, so that was one of the rare occasions where Vandrick and I argued regarding our son. Unfortunately, our son didn't stay for the party at Kennedy Park because I didn't want him to be disappointed. So, we spent some time hanging out together to take his mind off of missing out on his sister's birthday party. I disliked going to this particular place where it is always packed with kids, but that day I sucked up my feelings and took him to Chuck E. Cheese, and as their slogan stated, "Where a kid can be a kid!"

Later on, after he had calmed down, I talked with Vandrick and had him explain to me what his frustration was really about, aside from the fact that he wanted the kids to be dressed alike and at the party together. I found out the bigger issue had nothing to do with the clothing. Afterwards, he realized that no matter what our son had on, he should have stayed and enjoyed playing with his brothers and sisters. None of the kids cared about what they were wearing and by the end of the party they all were dirty with grass stains, cake and ice cream. We agreed we both played a role in the argument that took place and needed to learn how to handle ourselves better, especially in public settings. But we were learning, through trial and error, the balances of co-parenting even though we'd spent minimal amounts of time practicing it. So, after that call, I didn't even notice I hadn't heard from Vandrick in a while. I am so used to living my life and doing the things that I need to do; I don't spend much time thinking about things that were not a part of my everyday schedule. I was in my own relationship and wasn't focused on how much time had passed. I didn't realize months had gone by and not a single call had come in from Vandrick. Not a single word.

CHAPTER THREE

Unforeseen Circumstances

I took my car to have my smog emissions test completed at this well-known location that my boyfriend recommended. We were in the Fruitvale district of Oakland, waiting in line for my car to be the next to get the test. I got out of the vehicle and stood beside my boyfriend as he talked with the mechanics at the shop. There was a large black man ahead of us. His car was already going through the emission process. I remained standing in the same spot and I noticed he was looking at me as if we had met before, with a slight squint to his eyes. My boyfriend owned a nightclub and I frequented it regularly. I worked there, helping him out often enough to become a fixture in the place. I'm normally in the box office and on slower nights I actually get the opportunity to enjoy a few dance moves on the dance floor. I guess he could have known me from seeing me at the nightclub. This guy, looking at me as if he knew me, didn't spark my attention as anything unusual. However, it had the potential to make my boyfriend a bit uncomfortable because I was taken by surprise when this guy walked over and casually started to engage me in conversation. I asked myself, "*Did he not see my boyfriend right here? Did he not care?*" My thoughts were racing while I was trying to hurry and figure out how not to make my boyfriend uncomfortable before the situation became any more awkward than it already was.

"I apologize for staring, but you look so familiar to me," he said.

I hurriedly replied, "Yeah, I hear that often. I look just like my mother."

He replied, "No, I know you but I can't put my finger on it."

I was thinking to myself, hurry up and put your finger on it because my boyfriend was paying less attention to this mechanic and was now ear-hustling in on our conversation. "Well, maybe we went to school together?" I suggested. I was trying to help him reach a conclusion and hoping it was quick. Apparently, he didn't care one bit because he was taking his time to figure this out. He named a high school in East Oakland that many people had attended. I know that surely wasn't it because I didn't go to any schools that everyone in Oakland frequented so that was an immediate, "No," from me. I mentioned the club because my boyfriend was now sliding closer in on our conversation. I was trying to figure out who this was so that I could properly introduce this man that refused to stop talking to me as he appeared to rack his brain, trying to figure out who I was.

He responded, "I don't go out to the clubs often."

So that wasn't it either. Then I really looked at him and it was like a light bulb went off in his brain. His eyes widened and he started grinning. *Did I miss the joke?* I thought. Although I started smiling, it was from pure confusion and the awkwardness of the moment. He apparently figured it out, but I still didn't have a clue. This man did not look familiar to me at all.

He said, "I know where I know you from."

I replied, "Where?" I was extremely intrigued at this point. He named the block I grew up on. I laughed and said, "Yeah, okay. But I don't recognize who you are."

He said, "Let me tell you how much I know about you."

"Okay, tell me," I agreed.

"You have a baby by my boy, Boo," he remarked.

I looked at him and I was sure he could see the look on my face, a look that said, "Yeah, you know me." At that moment, hearing his name, it dawned on me how long it had actually been since I last heard anything from Vandrick.

"How's he doing?" he inquired.

I looked at him with a puzzled look on my face as I responded, "I don't know. I haven't heard from him lately."

He looked back at me with a surprised look and said, "Oh you don't know?"

I replied, "Know what?"

"He's in jail," he stated matter-of-factly.

I felt a sense of relief as he said that because my heart had clinched and my "oh shit" factor had gone up to 10 because I was thinking the worst. He could be dead, but jail was slightly more tolerable. Okay, that I could handle.

With a noticeable sigh I responded, "No, I didn't know but now it explains why it's been a minute since I last heard from him."

That, too, was uncommon for Vandrick in comparison to the past. *"Why wouldn't he at least have sent someone on the block to give me the message?"* This was my first thought because anybody that was at or near my family's old house would have known how to reach out to me, not to mention I had family that was still living on the block, too. I was not hard to find and he knew this. All of these plausible ways to contact me were circling around in my head.

Then he interrupted my thoughts with a Cliff Note to let me know I was going to have to locate him this time. He said, "Yeah, what he's in there for this time is serious."

I took that information and I knew that was code for I would have to rapidly try to get in contact with Vandrick.

He continued, "I think he's in Santa Rita."

That helped, but wherever he was I had enough information and I was going to find him. I needed to know what the heck he had gotten himself into and why he'd forgotten to do what he had always done in the past and tried to give me a heads up; even if I could only fuss and say, "Why the hell didn't you find me?" I thanked the man—I don't think I remember even getting his name—for the information. I looked over and felt my boyfriend had ear-hustled enough so I didn't have to replay the whole scenario again to him. In fact, I was sure of it. I said goodbye to the man I now know who went by the name of Dame.

"Tell Boo I said, what's up?" the man requested as he left.

"So, your son's father is in jail, huh?" my boyfriend asked me in a cynical tone.

I know he didn't care for him at all so I disregarded the question and gave a nod of my head to indicate it was not a subject I was about to entertain with him. My car was complete and passed the smog test. I was glad to hear that, but my mind is officially somewhere else. Later that evening, I didn't stay the night at my boyfriend's house. I was at home in my apartment where I could lounge and relax, but also perform my own investigation to see if Vandrick was still in jail. I called the sheriff technician's desk because they are the ones who answer the phones and provide inmate location information. I gave them Vandrick's name and date of birth. They confirmed for me that he was in custody then gave me his identifying information I would need if I wrote him a letter. After the phone call I did just that. I started to compose a letter to Vandrick to find out why he hadn't reached out to me and his son? Also, I wanted to make sure he had our contact information so that it was not a reason for the lack of communication. Apparently, he received the letter pretty quickly for snail mail.

It wasn't long before I received the first collect call from that 925-area code. Vandrick was happy that I reached out to him. He apologized for his delay in getting in contact with us, but he explained he had been going through a lot this time and was trying to tie up loose ends. He immediately asked me about the wellbeing of our son. I let him know he was doing well and I hadn't had any big trouble out of him. He was glad to hear that. Our son wasn't with me at the time so they couldn't talk over the phone. Vandrick asked that as soon as I could he hoped we would come and visit him. He preferred to talk in person and not over the phone. I assured him we would visit and it wasn't long before we made it up there to see him.

I found out the schedule and what days his building could have visitors so that we could fit them into our schedule. I let our son know we were going to see his dad. I'm sure he had lots of questions running through his mind because he had never heard this request from me. He didn't question it and went along with it. But, as we were pulling up to this huge facility with lots of fencing and barbed wire surrounding it, I got my first question from the back seat of the car.

"Mom, where are we going?" he asked curiously.

I replied, "Your dad is inside this place. We have to go inside in order to see him."

He simply responded, "Okay."

I know he really thought I was losing my mind as I was making sure he had no metal in his pockets or anything that would affect us getting past the metal detectors or the sheriff who stood guard, making sure every rule was followed. Again, our son didn't ask too many questions. When he does, I try my best to give him a semi-kid-friendly version about a very adult environment. The waiting process was about the worst part that we had to endure. That's when I was questioned multiple times with, "Mom, why is this taking so long?"

That was a bit harder to respond to because in kid's timing, five minutes was too long. I just tried to keep his curiosity at bay with the simple answer, "I don't know, but hopefully not much longer."

We were finally able to move through and pass the metal detectors, indicating our wait was almost over. We were walking hand-in- hand down a hallway with various doors and we could see other people sitting at little glass windows, holding phones and talking to an inmate as they looked at them through the small glass. Our son gripped my hand a little bit tighter. I walked with him and reassured him that he would be alright. I didn't want to be fussed at by any of the Officers, but I tried to let him see what I could see to familiarize him quickly to what was going on.

I told him, "You, your dad and I will be talking on the phone but this time you can see him through the little window."

Finally, we made it to where we would sit. It was in an area the size of a cubby-hole. I sat on the cold metal seat and put our son on my lap. Vandrick was already there, waiting for us to arrive. I picked up the phone, wiped it quickly with my sleeve, and I gave it to our son. They greeted each other and I could hear in both their responses this was not easy to adjust to, especially for a kid. He could talk to his dad and see him, but what about touch? He was used to giving his dad a hug. They talked for a while and I tried to give them their time to talk together, but everything was timed.

All too soon, Vandrick said, "I love you,"

Our son replied, "I love you, too."

I was handed the phone. He had still not informed me about what was going on, what the details were, but I do know the charges were extreme: attempted murder. He let me know that he was fighting the charges and was hopeful that they would have to let him go. They had no evidence to keep him locked up for what had taken place.

He tried to assure me as he said sternly, "Qiana, I was not there!"

I am limited to the questions I could ask, based upon the circumstances. Everything was recorded over the phone line and the Officers listened to every word you said in the hope of building a sure-fire, open and shut case. So, I can only look at him, nod my head and say, "Okay."

He let me know that his girlfriend, one of his younger kid's mothers, was soon to be his wife. They were getting married there at the jail and she would be able to keep everyone up to date on what was going on with him if we had questions. He assured me that she'd also be willing to participate with the kids in his absence. I was happy to hear that. I had always respected a person who got into a relationship where there were kids involved and they actively engaged with those children as well, and not felt as if their relationship was just about the two of them. And now he had six children, one of which was her own, she had a lot of responsibility to take on with all of them being so young. It would be somewhat easy on me because I navigated my household sufficiently and never required much from Vandrick. I didn't know what his soon-to-be wife would have to endure from the mothers of his other children and I didn't ask. Well, our visit was over far quicker than when it started. I held onto our son's hand and his grip was still tightly wrapped around mine. I understood because this place was not inviting or kid-friendly at all.

We made it outside and back to my car. Not much was spoken by my son until he was back inside the car and I could see his face in my rearview mirror.

I asked, "Hey, are you okay?"

The look on his face said he was not, but he picked up his toys in the back seat and he replied, "I'm okay," in a somber tone.

I found out later the visit had a negative effect on our son; he needed time to adjust back to his "normal" world. I would begin to realize this after a couple of visits, not that we went regularly because I hated taking him to that place. However, on the occasions that we did go, when he returned back to school, he would have a bad day.

One day when I picked him up from the after-school care program his teacher pulled me aside and asked, "Is he going through something? He's not his normal happy-go-lucky self."

I brushed her question off and replied, "Oh yeah, he just needs some time to himself on occasion."

I shy away from letting anyone know his dad was in jail. It made me uncomfortable because of the stereotypes that already existed in society. Even though it was not about me or my son and we had not been involved in any criminal activities, I found myself wearing the guilt and the shame of the crime Vandrick was charged with. I didn't feel comfortable discussing this topic unless it was absolutely necessary. I was learning this worked best for *me*. After a few more visits had passed, she asked again. This time he had gotten into an argument with one of the kids at the school, one of the kids we knew was considered to be his best friend. His behavior was escalating after the visits so I knew I had to speak up regardless of how uncomfortable it made me feel. I had to advocate for our son. I knew what the issue was and I knew how to address it. Contrary to what I wanted to do, I also understood if no one else knew it was only going to keep our son from dealing with his new reality. So, as the after-school program cleared out I stayed behind and we allowed my son to play as the teacher and I talked.

I broke my silence. "The days you have issues with him are right after I take him to visit his dad who is in jail.

He doesn't know how to process his feelings of not being able to have his normal interactions with his dad, so he wants to be left alone. He doesn't want anyone to bother him and force too much on him. He's trying to readjust after only being able to look at his dad through a glass window and talk to him on a phone."

His teacher wasn't familiar with having to be a kid going through something like this. First, she had no children and secondly, she likely had never been in a jail facility to visit anyone. However, she was sympathetic and empathetic to the circumstance I was bringing to her attention of a five-year old child having to visit someone he loved in a callous environment that couldn't adjust its protocols to accommodate children because of the risk of jeopardizing the safety of its facility. Therefore, it was easier for us to adjust for him in our daily interactions than to ask the jail to lighten up.

The teacher and I came up with a plan. I would let her know when we had a scheduled visit. When he returned back to school and the after-school program, she would have some solo-projects that he could work on. She would make the environment more adaptable for him to take his needed time to slowly acclimate back into his surroundings. This worked out well for all of us and after a few times of visiting we had mastered the plan. I am forever grateful that I spoke up and his teacher, Michelle, was so willing to provide a healthy and safe place for him at school. It was a complete success through my son's adjustment phase.

Well, that success didn't extend to Vandrick. He decided to continue and forge on into a full trial. I don't know if he was master-minding all the processes or if outside sources were coaching him in the decisions he needed to make. One thing I can say for sure was he never asked my advice and I figured Vandrick and those closest to him had it all under control. I assumed they knew at least a little something about the inner-workings of a trial. I assumed even if they didn't have a lot of knowledge, at minimum they would know his basic rights and if they were being infringed upon. I expected someone would speak up and make sure things went smoothly. I had been through a trial before with an old boyfriend and I remember all the work it took to assist him through it. I have had a love for law ever since I was a little girl. Anytime I was asked what I wanted to be when I grew up the answer would always be without hesitation, "A lawyer."

I took classes to make sure it was a passion of mine and I was certain it was what I wanted to do and who I wanted to become. It was only derailed when I decided boys and relationships were more important and I started to place that above my own aspirations. How easy we could shelve our dreams with just a few quick decisions. In that previous trial I assisted with, I was taking a business law class at the same time. And in lieu of going to sit in the classroom every day I negotiated with my instructor to sit through my boyfriend's trial instead.

My instructor told me, "What you are about to see during this trial will give you far more knowledge than anything I can teach you in a semester from a book."

We had a deal! The only time I had to be in class was on days that court was not in session and it was mandatory that I attend his class for my final exam which was a mock trial. Although I assisted where I could, what was most important to me was finding him adequate representation. During my second period keyboarding class I composed a mail-merge document that had every criminal attorney's mailing address listed in the Bay Area, pleading for them to take his case pro bono. Yes, I learned the meaning of the word pro bono quickly being that I was a teenager and definitely had no access to Johnnie Cochran, Jr. Esquire's type retainer fees. I had learned many things to gain knowledge in the subject of law first hand. Then I had to write a report for my business law instructor from my experiences during the trial which was how he'd determine if I passed or failed his course.

I can proudly say I participated as the defense attorney in the mock trial who gave the closing argument and the verdict was in our favor. I lived up to my end of the agreement between me and my instructor despite the misfortune of my ex-boyfriend's decisions that got him caught up in some criminal activity. I earned myself an A+ in my business law class. My ex-boyfriend was found guilty in his trial which we knew was the case from the moment he turned himself in. But, with all of our efforts he didn't face the most severe punishment possible. He was sentenced as a juvenile instead of as an adult. Therefore, he would not serve time in an adult facility. We found gratitude with the lesser sentence delivered.

As for Vandrick, he persisted in his innocence, putting his trust in the physical evidence he felt would exonerate him of the charges. He believed if his representing attorney would put on a proper defense, he would be acquitted and this cloud looming over his head would disappear. There were several extenuating circumstances that would prevent this from even being a possibility or heard in trial by the actual jury. Unfortunately, during his time in jail, he lost his mother. Any money he had at the time which he planned to use to pay for a private attorney was shifted to pay for funeral costs, along with bail for his big brother who was also sitting in jail in a different county. He wanted to make sure that they could attend their mother's funeral, no matter what. She never wanted her children to be in the situations they had gotten themselves into, but now they needed to be there for her and at least honor her in any way they could after her untimely death. So, less concerned about himself and more about his mother, during court proceedings he requested that he be allowed to attend her funeral. He was grateful the judge agreed to allow him to be escorted to the funeral by the jail's officers. There was a fee for this service, but he was able to pay it, and he took care of it immediately. That brought him some relief as he grieved.

He would be able to attend his mother's funeral, but the jail's sheriffs had a different plan altogether. As he awoke the morning of his mother's service, he prepared to attend her funeral but no officers came to get him from his cell. And later, they dismissed the approval of the judge by simply stating, "We weren't prepared for your movement." Despite the grief of losing your sole-provider, the matriarch, the only mother you were ever granted permission to love, the sheriffs would still find a way to play with your emotions by showing you they ran their own house. He mourned her death in his cell, reading with aching pain in his heart as his eyes filled with tears, the last letter he had received from his mother right before her death. The funeral service would go on with family, friends and associates in attendance while he did his best to have his own private memorial service.

Now, without money for a private attorney, his case has been transferred over to a less individualized public defender who'd only visited Vandrick a handful of times to put together partial pieces of his case. He'd complain about how he had multiple cases, not enough time and even fewer resources he was willing to dedicate solely to one defendant. His public defender's persuasiveness was leaning heavily toward getting his defendant to take a deal so that he could wrap up the case and move on to the next one in his pile of many. A trial was far less appealing than he wanted to go through, especially when Vandrick was unwilling to confess information that would affect his own safety and security. When you are actively entrenched in street life and appear to be helping out cops, or any part of law enforcement, you were basically writing your own death certificate, even by the hands of people who you had been loyal to. He was hoping things like DNA and forensic evidence would help to clear his name. However, it would never make it into his trial. His trial would be streamlined and focused, less about the facts and more about his reputation and activities in the streets, with the prosecution being handed the easy win.

Guilty! That was the verdict after already dealing with the devastation of his mother's death. Vandrick could hold his head high at the contribution he was able to provide toward the funeral service and burial for his mother, even if it meant losing his case in the process. He would battle with this outcome and use it as fuel to keep himself moving forward. Sentencing was the next phase in the process because changing the guilty verdict was something that would be postponed and dealt with in the appeals court. From the look of things, they wanted to throw the book at him. He was facing the possibility of 40 years to life.

I heard from Vandrick again through another collect call from the jailhouse. It had been a few days since I last received a call. He was still very numb and the words he spoke came across the phone line at a slow pace, as if he was reading them from a document versus words he was reciting from memory. I don't hold him accountable for this or even mention it because I could only imagine my condition if my whole world was shattering around me.

I asked him, "What happened during sentencing?"

"They sentenced me to…. 25 to life," he replied slowly.

I think as he was saying it, he was also telling it to himself. He was still in disbelief. I had no response. Trust me, there was nothing I could say that would remotely sound right in that moment. I was hoping something would come to mind to change the subject, but even that wasn't working so the phone was silent for a period of time. It eventually came to me. I talked about our son. I let him know that he was doing well in school and other small talk just to make the conversation light. I did ask about him transferring to a state facility, but he didn't know the exact timeframe. However, soon he would be moving over there. One thing we did agree about was that over there he would be able to have contact visits. Therefore, there would be no more cold and emotionless visits through that dreary glass for our son, who at this point disliked going to visit and I was not going to force him to endure that place if he no longer wanted to.

CHAPTER FOUR

Prison Principles

I had missed him! He had already left the Santa Rita County Jail. I guess I wrote him too late as I looked down at the envelope, I'd just taken from my mailbox which was addressed to Vandrick but stamped, "UNDELIVERABLE - RETURN TO SENDER" with a handwritten note on the envelope which read, "No longer at this facility." The transition had taken place. He'd been transferred to the reception center. No more thinly wired glass windows, telephones with bad connections, and cold metal seats that felt more like you were sitting on a pole versus a stool. Don't move too much left or right or you might end up on the floor. What still remained was that no matter where he went there was the lack of privacy, barbed wire fencing, officers now referred to as Correctional Officers (or C.O.'s), restrictions, long lines of waiting, and on any given day a whole lot of confusion and dysfunction.

Vandrick was finally able to make a call. The collect call number had changed and it was no longer that 925-area code. But the recording was basically the same, "You have a collect call from 'Boo'..." Now that he'd made it into a state of California facility, he let me know he mailed out visitor applications that needed to be completed and returned to the prison before we could visit. They wanted to know about me and our son. I didn't like that part. I am the same girl from the neighborhood with the same mentality, "That's for me to know and you to try to figure out." They don't need to know about me. I am just a person who had a kid with him. "*What did all my personal information have to do with him? Why did I have to apply to visit? Don't they know I was really not that invested in any of this? I was more like a back-up driver that made sure our son got the opportunity to see his dad. Not to mention I was pregnant and going through a serious complication. I was neither emotionally nor physically equipped to deal with much outside of what I was going through.*" These were all the thoughts running through my mind during this call.

Vandrick knew my tolerance level was minimal, but he actually asked me to notify his counselor about my hardship as it might assist him in not being moved too far away. He said I could call his wife and get the needed information from her to complete this process. I already knew I wasn't disclosing my medical information to these people. This was far too invasive for me and the subject was too sensitive for me to discuss beyond my immediate family. I considered Vandrick as family and that was the only reason he knew. After receiving the returned letter in the mail, I found out his information and wrote to him at his facility. That was how he even knew what was going on with me. We spoke a little bit longer as the phone calls were timed. I made sure he was able to talk to our son for the remainder of the call.

Out there he called far less than he did at the county jail. I don't know why, but there was a big difference for sure. Visits and writing were apparently the preferred method of communication in California State Prisons. However, he did call before and after when I finally had my surgery. I appreciated his concern. And he called when I was approved to visit to let me know that the process was complete. Yes, I had to let the state into my business for the sake of our son. I didn't visit for a long time, though. I had to get through recovery, the birth of my daughter and then build up the motivation to go to a state facility. Even the thought process of preparing to go was intimidating. When his wife would visit, she would take the kids with her most of the time, so that also cut down on me having to make trips up there. But soon enough, he made a request for me to come and bring our son.

The first time I went to visit he was about 150 miles away from where I lived. That was better than where he was before. They had moved him at one point in time so close to Mexico that you might as well have considered the facility there instead of the United States. I loved to drive so 150 miles was not bad at all. But let me be clear, I loved to drive scenic routes and to get to desirable destinations, not locked inside of a correctional facility. In preparation, I had to make sure I pumped enough milk the day before for my baby who would be staying home with her father. I had to also take down my hair because according to the Title 15 regulations, it was not permitted for me to wear specific hairstyles and I already knew I was not getting into any facility with the number of bobby pins I had in my braided hairstyle.

I made sure I had no underwire in my bra. I checked, changed and talked myself out of five different outfits because they might not pass the CO's discretion which was only based on the mood they were in on that day. I must remember to not wear the color blue. I made sure I had my driver's license and our son's birth certificate in a clear plastic bag. I went to the bank the day prior because the only denomination of money that could be brought into the facility were $1 dollar bills or quarters. I dressed our son in clothes which were either black and white or just black because they had no problem with not letting kids inside based on their clothing as well. Then I prepared and packed my vehicle for the drive: gas (check), extra clothes (check), a small cooler with food and an energy drink (check); games for our son (check), a blanket and pillow for him as well (check) and directions to the facility (check). I was starting to think people prepared and packed less for war. I'm sure I forgot something, but I hoped it was nothing super important or we would just leave and make that two-and-a-half-hour drive back home, apologizing when he called and asked why we didn't make it inside. I had already been warned by his wife to be there by 6:00 a.m. so that I'm early enough to receive a visitor's pass because visits were based on a first come-first served basis.

So, to make sure I correctly calculated for any bathroom breaks or unforeseen situations, we planned to hit the road by 3:15 a.m. This was definitely torture for the children being dragged through this process, including my own son. This was now his new reality in order to maintain some type of relationship with his dad. It was still pitch-black outside. Only my music, my thoughts and feelings of obligation kept me entertained as I drove down Highway 101 South with my son sleeping from us being up so early in the morning. I wondered if the powers that be, the judges handing down those prison sentences, ever thought about the fact that the inmate was not the only person being sentenced during those trials. The entire family was now being convicted of a crime that they didn't participate in.

The sun had started peering over the distant hills as I got closer to Salinas Valley State Prison. Our son was slowly waking up, opening his eyes and blinking constantly as he realized the car was still rolling down the highway.

"Are we almost there?" he asked in a sleepy voice.

"Good morning to you, too," I teased him. "Yes, we are not too far away."

Shortly thereafter, as we exited the freeway, we were greeted with the most dangerous scene ever as cars were lined up right off the exit of the freeway. I'd hate for someone to come off this exit too fast for whatever reason. It was a fatal multi-car accident just waiting to happen. Well, we had no choice but to line up off this exit as well and hopefully they would let us drive onto the facility grounds before something terrible happened. Our son grabbed a snack and his electronic game to entertain himself as we waited. Before long the officer arrived and flagged us to proceed toward the entrance of the facility to receive our visitor's pass. Our number was less than 50 so we would make it in during the morning group. We parked and got out of the car and went inside this little trailer house near the entrance to use the bathroom. I needed to throw a little water on my face as a quick pick-me-up. I looked around a bit and saw they had a television, clothing for people who needed to change, toys for kids and a kitchenette, but I don't know if that is more for the staff or for the visitors. I didn't ask as I got my son who was sitting on the couch, watching the television. We walked back to the car to drive the rest of the way up the hill to where the facility was located.

On the grounds were two State of California prisons — Soledad and Salinas Valley. I had to make sure I was going to the right one. Once we made it to the facility, this was where the visiting process would begin and I would find out if I had everything we needed and if we were properly dressed. We were called into the processing office in groups of 20, about every 30 minutes to an hour, for processing. If I did everything correctly, we would be cleared and free to go through the metal detectors and on to visiting Vandrick.

As the group of us walked down the road to get to the visiting units, we could hear frantic tapping on the windows from inmates trying to gain our attention. Those sounds of desperation entered my mind. If no one ever came to see me in a place like this, I might have resorted to tapping on the windows when I saw people that were not in uniform and had no authority over my life as well. But I did not turn my head at all to see where the sounds were coming from out of fear that some Officer in the distance might see my movements and I would be yelled at and humiliated for something as simple as a reflex reaction to look. I continued to walk towards the visiting room, face forward, along with the group of other visitors to where the actual visit would take place.

This was very different from the visits at the county jail. There were knee level tables and chairs that reminded me of kindergarten classroom furniture. I was thinking, *"Were they not prepared for full grown adults to come and visit? What is a 300+ pound visitor or inmate going to do with this little chair?"* I'd hate for someone with long legs to try to sit with their limbs resting uncomfortably against this table. Everything was spread apart and there was a podium in front with two Correctional Officers present.

There was a place for visits behind the glass, but thankfully we were not here for that. Sorry, but Vandrick would have to be mad because I was not driving all this way to sit and look at him while we talked over the phone. From what I had witnessed, those visits seemed to conclude quickly.

We had done the glass and phone thing already and I was not ready to drive two and a half hours back home with a despondent little boy whose feelings would be scattered with confusion as to why he couldn't at least hug his dad. I prayed Vandrick never placed that burden on me to let him down gradually as I explained to him why. —I am certain I would not participate in the behind-the-glass visits. We took our seats in those kindergarten chairs as we waited, watching the door at the back of the room close and open as inmates came in and greeted their visitors. I looked at our son and he was not as nervous as he used to be. It had been a few years of dealing with this jailhouse stuff, so I guess he had become more resilient to the sterility of the environment.

But I still asked him, "Hey, are you okay?"

He looked at me as he replied, "Umm-hmm, yeah."

Finally, the door opened and we both saw a familiar face. A tall, thick frame was filling the doorway. He walked into the room and the door slammed behind him.

The sound was a definite, "SLAM! CLACK!" as if shouting, "Either you're in or you're out!" Vandrick walked toward our table and our son got up to greet his dad with a big bear hug. I had a grin on my face as it was apparent no matter what the circumstances were, their bond was genuine. They both needed that hug. Our son had never been a small, frail child as some kids were who grew out of their baby-fat and thinned right on out. That was not the case for our son. His frame was solid and full just like his dad's. He may have thinned out a little but it was not extremely noticeable. Both had big heads, broad shoulders and solid chests. Their complexions were close in comparison and their facial features were exact—squinted eyes that closed when they smiled. They had cheeks which filled their faces. It was amazing to me how he was carried full term by me, but looked just like his dad.

I grinned again as I also noticed Vandrick had not sat down and was looking at me. Oh, he was waiting for a proper greeting. It was awkward, initially, because we were always cordial and friends no doubt, but we had never had to establish certain romantic etiquettes. I stood and I walked towards him and being funny I extended my hand to shake his. He looked at it and laughed, "You've gotta be kidding." I laughed, too, as he grabbed my hand and pulled me in for a big ol' bear hug similar to what I'd just witnessed between father and son. I was almost a foot shorter than Vandrick, and trust me, it looked just like their hug. I was swallowed up by his large stature. Afterwards, he was finally comfortable with sitting down in a kindergarten chair, adjusting as best he could to make it more comfortable to sit throughout the visit. We all talked, laughed and even had moments of tears; well I did. The guys wouldn't admit they wept a bit, too. There were no full-on tears requiring napkins to absorb, but no one could deny that there wouldn't be any emotions after all the time that had passed. They walked around and caught up while I sat and took in the surroundings of the room. I realized that many people didn't move around in this environment. Then I noticed lines on the floor and signs posted which read, "Out of Bounds" and messages that reminded me quickly that whatever was done was heavily monitored so that control could be maintained.

They came back to the table and were ready to eat. I knew both of them and food were not separated for too long. I grabbed the clear bag that had the money and headed towards the vending machines. They made their selections and I didn't even have to question the fact that they would somehow incorporate chicken into the menu. They loved it over all other choices. They chose some other things and I took it all over to the bank of microwaves. I looked down and I was standing on a red line with white writing that said, "Out of Bounds." I looked at it intensely then looked back at Vandrick. He saw my look of curiosity and said, "No, that warning is for me." I stepped across it and warmed their food as I observed another sign near the vending machines stating, "Inmates cannot handle money." Wow, they had gone from freedom, street status and having material things to being humiliated and disgraced. How quickly the tables were turned once you were in this place. All the things these young people had been fighting, killing, stealing and gaining from the streets were immediately stripped from them.

Once their food was ready, we returned to our seats. I don't get much of anything for myself because I know I was the only designated driver and I didn't want to eat too much and then become sleepy out on the road. I was carrying precious cargo and we wanted to make it home safely. I could always grab something once I was closer to the Bay Area; I was sure our son would prefer that as well rather than what was in those vending machines. Those were more like snacks and less like meals, but I guess if you ate plenty it would substitute for a hearty meal. Plus, for the inmates, I assumed it was definitely a better alternative than what they were serving on the other side of that door.

We talked a little bit more. I was finally able to ask a couple questions about his case, not a lot since the visit was for them, but I just wanted to make sure he was doing okay and still fighting for his freedom. He assured me he was working on trying to get out of this place. As he put it, "This is no place for a man." He made sure it was heard by our son to remind him not to ever do anything that would bring him inside this cement. He described for him the disgust that would be waiting for him on the other side of that sealed door: constant fighting and lives being taken by one another; men in rooms with each other in bunks one on top of the other; toilets just steps away from their heads where privacy was not an option. Humiliation was the key to breaking a man's spirit, whether by staff or other inmates. Prison politics was a real thing. If you broke a rule, you would face the consequences and so-called apologies, saying, "I didn't know," is not a way out of the predetermined punishment. He reminded our son, "It is so much easier to listen to your mom. She is not telling you to do things like your homework, clean up your room and to respect your elders in order to be mean. Obey her requests because she is only trying to raise you as best she can and keep you from being in here." He added, "I wish I would've listened to her and my mother as well."

I looked up at him. I wondered if he was just saying this or if he remembered the night that we spent together in the motel room when he had picked me up from my sister's apartment in San Leandro, CA. Later that evening I found out that he had been shot with buckshot and his back was riddled with tiny holes. I placed damp towels on his back and told him he needed to go see a doctor. I was scared for him as he described the incident that led to this outcome. After hearing the details I pleaded as best I could for him to give up the streets. I knew he had potential beyond hanging out and making fast money. He could become anything he put his mind to. He was extremely intelligent and beyond the bad boy image he had to maintain he had a very caring heart. He was the type of person that would put other people's needs above his own. Well, in a way, I guess that also explained how he ended up in this place—putting other people's needs above his own, yet again! This was what my grandma meant when she said, "Loyal to a fault." I have this same quality, but I thank God, it's never landed me in prison!

A few hours had passed and just as I was getting sleepy and hoping for a second wind of energy to come over me, the visit was coming to an end. This wasn't nearly as bad as the county jail. Here, they got to interact with each other for hours, not minutes, and they were able to touch. I think we take for granted the human touch, especially from one loved one to another. That connection is so powerful. I remember seeing research on it before; a baby's response when it was not held by its mother for an extended period of time versus a baby's response who was cuddled and had bonded with its mom. Human touch and human interaction are far more effective than we give credit to. However, I think the jail systems know this, too, so they are using it counter-productively with inmates to break them down. Yeah, my thoughts are often deep. Some may think I was a conspiracy theorist if they took a tour through my mind, but I considered it paying closer attention than most. As the saying goes, "Beauty is in the eye of the beholder." However, I can change it to suit me: "Logic is in the mind of the beholder."

We're now being dismissed by the officers as they reminded all of the inmates and visitors that were allowed "one 'brief' embrace as you part from your visitors." We walked toward the exit where it was a bittersweet moment as I witnessed the father and son bear-hug again. This time was slightly different as I noticed their hug lasted a bit longer and there was a sense that they didn't want this moment to end. However, they did release each other and also smiled at one another as Vandrick told him, "Boy, be good," and our son nodded in agreement, like he's so cool. It was my turn as I looked at him again and extended my hand. I found this gesture funny as I tried it a second time. He knew we didn't get down like that with mushy stuff, but again he pulled me in by my extended hand for a hug and this time he whispered into my ear, "Thank you! Thank you for bringing my son to see me."

I responded, "You're welcome. That's what friends are for." I don't know if he could hear me because I was kind of wrapped up in the middle of his chest from being so small. This time I got to measure us and the top of my head was right below his shoulder when we were standing up straight. Yeah, I am much shorter than he. I never noticed before and I would likely never openly admit it. I took in the moment and the visit was over. We processed out, passing through the sally-ports as people were asking questions of the CO's staff at the front desk. I just wanted to get to my car and start my journey on that long drive home. Any questions I might have had were far less important than me getting out of this place.

I made it to my car and we made one last stop before exiting the gate to use the bathroom again at the family resource center that was onsite in that little trailer. We moved quickly because I wanted to make it home before it got totally dark. We had spent the entire day visiting and I could honestly say I was certain I had been awake over 12 hours already with little sleep the night before just trying to prepare so we didn't forget a single thing. I knew even if I didn't mentally accept it and feed into it, my body would soon be telling me, "You're tired." We got back into the car and headed onto the highway. I had my energy drink in the cup holder ready to start sipping on it as I traveled towards our destination. I think we'd made it about two exits before my cell phone started ringing. A robotic voice said, "You have a collect call from, Boo."

"I accept the call. Hey, what's going on?" I asked hurriedly as I thought maybe something may have happened.

He replied, "Hey, nothing. Just making sure you got on the road okay. Plus, I saw you were getting a bit sleepy."

"Umm hmm," I agreed, "but I'm doing okay right now. I'm drinking my energy drink and once I hit the fresh air, I had a burst of energy. I'm trying to get as close to the Bay Area as I can before tiredness starts to wear on me completely." He was quietly listening and I don't know if he was paying attention to me or looking at things going on around him.

He finally said, "Oh okay. Qiana, thank you, again, for bringing him to see me. I needed that. I hope this won't be a rare occasion. I enjoyed you visiting as well."

"You're welcome," I replied, keeping it all business. He asked about our son. I let him know he enjoyed himself and was already sleeping. This experience was less traumatizing for him than the experiences of going to the county jail. However, I would prefer not to ever have to go back to any of them whether it was a county facility or state. He reminded me to be safe on the road as our call came to an end. I turned the radio volume back up and continued my drive home to reunite with my eagerly waiting daughter who hadn't snuggled with mommy since about two o'clock in the morning. And I'm so tired I'll have to spend the night at her dad's house instead of in my bed at home.

We made it to his house safely. I paid close attention to our son and how he would adjust to home and school life after visiting. This time was a smoother adjustment with no detachment from people and wanting his protective isolation. I needed to see that so I'd know whether or not this was going to become too much for him or he'd be okay. If there was an issue, Vandrick was just going to have to understand my need to protect our son and why there would be a lapse of time necessary in between visits. Thankfully, that wasn't the case. My son adjusted as smoothly as could be expected in returning back to his normal routines as well as playing and interacting with his friends as he always had.

It would be a while before I would make the drive again. I had other things that I was focusing on during that time that took priority over driving 150 miles down the road to make sure we were at visit. Plus, his wife made sure that the kids were able to visit and it was much better when they all visited together. Then it wouldn't be just one kid sitting and listening to a bunch of grown-up talk. They could entertain each other. There were games and toys as well, so it made it easier to find things for them to do. My priorities had shifted because I was working full time for the County of Alameda as I had been for a few years now. I had taken on a higher paid position which came with more responsibilities, not to mention I had started focusing on going back to school full-time as I was adjusting my career path into becoming a nurse. This was not my girlish dream at all of becoming a lawyer, but it was a safe option towards a fulfilling career. I loved to help people so why not help them medically. And after the complications of my pregnancy, my knowledge in this field definitely would be therapeutic for me as well.

Now, with the responsibility of two children instead of just one, my time became increasingly more limited. As I had witnessed before, no matter how great a plan is made, some things might need to change. I am no longer in a relationship with my daughter's father. I am again doing what I know to do best—take care of my kids and press through to make a life better for myself and them. We are amicable and I can proudly say he was participating in helping to raise our daughter as well. He loved and adored his child. I jokingly said to him, "She is your best friend." He couldn't do anything but agree. When she was in his presence, nothing else mattered. He was older than I was and this was his first biological child. So I expected his attachment to her to be strong.

I still had the same view as I had before. I would never interfere or intervene in their relationship regardless of whether or not our relationship soared or flopped. I hated to see children used as pawns for some frivolous hope of hurting the other parent based on failed relationships. That was not what our children were here for and I also believed in karma, so what you do to others will ultimately come back to bite you in the butt 10 times harder.

CHAPTER FIVE

Traveling the Distance

Along with all the things I had going on in my life, I had moved to a very nice apartment complex outside of the vastly different Oakland, CA. I needed to expose my children to a different environment where they could play peacefully outside and be given the option, versus a requirement, to be in the house playing. There would be no more reckless drivers to dodge, random shootings and bullets with no name attached, drug and alcohol paraphernalia left right outside the doorstep. These were influences that were hard to pull kid's mind from as they witnessed their parent's daily struggle while the drug dealers had all the nice material things and parents preached and gave lectures of, "Work hard, go to school, get good grades, and you will be successful," which would fall on deaf ears. I had a young black male and female and I needed them to see something different and to let them know that success comes in different formats. I wanted them to know I was okay with them coloring outside of the lines if it makes them happy, and that the sky is really the limit.

It was difficult for me, moving out this way because financially it does cost more, but I would not give in to that. I knew sacrifice and struggle would come with my decision and I was willing to go through that for my children. They were in a good educational district. They wouldn't need to use a friend or family member's address in order to attend those schools as my sister and I had gone through. It was a shame that we still had what was known as "good" and "bad" school systems. But it was the absolute truth. I knew some children who went to school and got straight A's because the teacher was relieved by the fact that the child came to class and was not a problem for them; not because they were able to determine their educational readiness, or the assignments were so easy that they deserved an A if they could complete an answer in a minute or less.

I am always more appreciative when my children are being challenged academically. I needed them to test their ability to problem solve and not just have the answers given to them. They would find as they grew that the world would not give them anything that they did not earn, and they would be faced with challenges. I needed them to be prepared so that they didn't fold from the pressure. Technology was equally important and the schools where we moved had total access to it. There was not a classroom at any of those schools that didn't have computers, laptops, or notebooks for each of their students, including ones they could borrow and take home if for some reason they didn't have one, or the one at home was not working.

I recalled when I first walked into my son's cooking class for Back-to-School Night. I was floored as I thought back to my own home economics class and what it looked like in comparison to his classroom. Each cooking station, six of them to be exact, had its own sink, microwave, stove and matching pots and pans set that I quickly identified had come from Costco. I was hoping his teacher didn't see my complete amazement. I was sure I was likely sitting there with my mouth wide open. In my class there had been one stove that the entire class would share and I thought I was attending a good school, or at least better than the schools in my neighborhood, but these schools here were even better than those. Every classroom was as sophisticated as any four-year college classroom. Those schools were preparing these kids not to be intimidated by the future. They were introducing them to it from childhood on up. For this I would make every available sacrifice to live in the city of San Ramon, California and call it home.

During the times I could not visit, Vandrick and my mode of communication were phone calls and letters. He was patient with me when I was unable to show up physically. He understood that I was building and transforming my life to make things better for me and my children. Even if it turned out not to be a complete success, it wasn't going to be because I didn't put forth every effort. One thing we have gained over this time was the ability to communicate with each other. He was becoming more of a friend than I ever recognized in him before. We began to talk to each other about everything; not much about his case but more about our personal lives. He knew what was going on with me and I knew what was going on with him. If I needed a male opinion, I was not afraid to ask him, and likewise if he needed a female's perspective, he didn't hesitate to ask me.

We began to weigh our friendships as well. Along the path of growth and development we had to rid our lives of the dead weight of some people. And he was excellent at being blunt about people I needed to let go of that I was allowing to consume my space. He was also learning that as well. So, friendships that no longer served him great purpose were being severed. What his life was mirroring today involved people who loved him on the streets only for what he could give them, but now they were not answering calls, changing phone numbers, unwilling to help him out financially, or even non-financially. He was seeing people for who they truly were. With material items and financial access to him, they all appeared to be friends, but without those things they would eventually disappear. Those same people were now living their lives and had forgotten about his life that continued behind that concrete wall. If he wasn't reaching out to them to remain relevant in their life, they surely weren't reaching inside to remain present in his. He was learning and would continue to learn, "Out of sight, out of mind."

I would eventually return to visiting him through all the obstacles and as prepared as I had been before. The nervous energy of making sure I got everything right the night before still bugged me. There was the late night up prior to the visit day and the blaring of the alarm in the wee hours of the morning for me to let go of my pillow and get out of my bed. Then I needed to wake my son up to make sure he put on the outfit that I picked out for him to wear and get him to wash his face and brush his teeth. I don't fuss with him too much because as soon as we get in the car, he falls right back asleep. I took my daughter to her dad's house the night before so that I didn't have to try to get her there at such an early hour before heading down the highway to visit Vandrick. Thank goodness she was no longer exclusively breastfeeding so her dad could prepare her meals without needing me.

Before long we are in the car and on Highway 680 South as we start our journey to see Vandrick who was still at the same facility. Everything was pretty much the same—the same long dark roads and me listening to music and spending this much needed time to think about nothing in particular; it was nice when you had the time to just sort your thoughts out. That was what driving did for me. It was my time to allow my thoughts to come in and the ones I no longer needed I would gradually dissect and let them escape into the darkness of night. As we got closer to the facility the sun rose as it always had before. That was definitely God's paintbrush as I watched the colors of the sky transition from night to day.

I looked at all the fields as I traveled along the highway and I thought about the field workers—the majority were immigrants—that had what appeared to be endless amounts of work. Those fields were massive and couldn't be tended to with only machinery. Recently there had been news about a field worker that died due to heat exhaustion which made my thoughts focus there. I was looking to see if there was even a shelter for the workers to find refuge from the beating sun. I looked for water sources so they could cool down if there was a need. I had never worked in a field such as this, but these thoughts came to mind as I was taking in the scenery.

Soon, I would exit and line up again in this unsafe location. Some of the cars were very familiar. I remembered them from the visit before. There was this one car that stood out the most—it was car number one this time and it had been car number one the last time. It had to be the woman that I had heard about who visited here every weekend to see her husband. She would get here on a Friday night and sleep on the side of the road in her vehicle pretty much until Sunday when her two-day visit was finally over. I considered that extreme dedication and a love that I had yet to know to that capacity. I did not judge her at all. I prayed for her safety. I did not visit here regularly and if I knew about her, I know others did as well. I prayed that she was never hurt in an attempt to spend those hours with her husband. I am certain that would be devastating for him to endure. The guilt would eat him up from the inside out. She was only honoring her obligations as his wife and doing what she could to support him through this trial in their marriage. I prayed she was protected and God continued to grant her traveling grace and someday this test of faith would permanently end for her.

Finally, the Gate Officer had arrived and we were slowly driving onto the facility grounds to receive our visitors' passes. Again, we were ushered into the Center to freshen up a bit and relieve our bladders. We had a small routine we followed. Then we returned to our cars and traveled up to our designated facility. I saw some of the same faces as I got out the car to see what numbers they had called, if any. While I was outside a lady near me looked down at my feet.

"You are not going to be able to get in with those shoes. They require that your shoes have a strap around the back," she said to me.

I looked down and thought to myself, *"Oh my goodness, I messed up this part of my process of 'getting it right' and I surely don't want them to say anything to me."*

The lady added, "There is a Payless Shoe Source store that will be opening soon just down the highway a bit.

I've had to go there before so I know they should be open by the time you get there."

I thanked her and headed back to the car in a hurry and got in. Our son looked up at me from playing with his game and asked, "Where are we going?"

I said to him with a sigh, "I have to go buy shoes. I wore the wrong ones." We left the facility and headed north on 101 just a few miles back and there was a shopping center off the exit, and just as the lady had said there was a Payless Shoe Source there. But more importantly, it was open! I rushed inside and the cashier saw I seemed to be in a hurry, so she asked, "Is there anything in particular that you are looking for?"

I said "Anything that is of a similar color to my blouse, but enclosed with a strap is the most important thing."

The cashier giggled a little and replied, "You must be going to the prison."

I said, "Yes."

"No worries" she assured me," we get this request a lot. It is hot out here so many people want to wear a sandal, but it has to be a specific sandal in order to get into the prison."

We found a shoe that would work. I purchased it then hurried back up to the prison. As I arrived back on the grounds where processing was taking place, I asked a visitor I saw standing around what number they were on? They told me and I was relieved they had not called our number yet, but they were getting close. I changed my shirt just in case it was too transparent—a button down white and pink blouse—but my nerves were bad from the last wardrobe malfunction. So, I put on a rust-colored shirt with caramel-colored stripes. I no longer cared that my pink shoes might not be an exact match. It was close enough and I was not there for a date. This was a visit to a prison for the benefit of my son to see his father, so thoughts of colors clashing would need to subside.

There were no glitches with anything else and we had another successful visit. It was all about the same. I did get my first chance to see how a couple would get married at the facility. It was done during visitation. The facility had a room off to the side of the visiting room and the couple and their witnesses went in there with the preacher and not too long afterwards they were married. As the bride and groom returned into the main area of the visiting room, they were greeted and congratulated on their nuptials by inmates and visitors who knew them. For one moment in time everything was festive.

Another visit was complete and we made it home safely. It was a nice time spent together as we caught up on a few things. Vandrick had uninterrupted time to spend with his son and occasional small talk with me. It dawned on me that we didn't discuss his case much at all, but I hope for his sake that all is well. I hadn't heard any of the details that resulted in him being incarcerated or even why they punished him so severely. Whatever it was, I did wonder how he was holding up? He never complained or sought pity with the constant grumblings of "woe is me." I'm sure many inmates would indulge in it if they felt they were unjustly sentenced to a crime they didn't commit. Instead, all I knew was that he said he was not guilty of the crime and he felt they prosecuted him for not cooperating by identifying the actual participants. Maybe this was God's way of gaining his attention. I had witnessed some subtle changes in Vandrick, but I only noticed them and never brought them up. I didn't know if it was time taking a toll on him, or if he was just deciding to not fight maturity. Now when we speak, he has a sense of wisdom in his opinions. I have always acknowledged his intelligence, but I was seeing him implement it more.

Vandrick did as he had done in the past and called to check on us as we hit the road. We were well on our way and headed for home to eat and relax. Visiting was a process that definitely took a toll on me by the end of the day. He again thanked me for making provisions for our son to see him. All I could do was acknowledge his gratitude with a simple, "You're welcome." I was doing as I knew I would want anyone to do for me if the shoe was on the other foot and I was the person incarcerated. I realized no one was exempt from the possibility of being put into prison. I wouldn't want to be forgotten about and left alone as if no one actually cared. As people tell you they care but it's not supported by their actions. The reality of their absence speaks greater volumes than saying the words, "I love you."

I made sure after each visit to evaluate how my son readjusts back to his normal routines just as I had in the past. He was my priority and major concern. The determining factor of when and how we would make the next visit was whether or not he could handle the last. If he was adjusting well and this was not having intense side effects then we would maintain the visits for the sake of their father-son relationship. But if not, we took the breaks needed to stabilize his mentality and moods throughout this process. Prison was not a place for kids and it had been proven that the slightest thing could have lasting effects. I appreciated that the visits with Vandrick were not solely moments for him to get out of his cell to eat, but a time that he spent trying to do everything he could to contribute to the growth and development of our son.

He seemed to have learned one thing for certain; he did not want his children to walk even a mile in his shoes. He reminded our son at every visit, "If you ever come in here for some crime you have committed, I am not going to make prison easy for you and have you bunking with me. I have already told you that I don't want you coming into prison." I know this was a mere threat and if any unfortunate event happened as a result of any of his children's actions, he would do everything in his power to protect them from the gruesome realities of prison. However, it was a threat that he hoped would resonate in their minds and actions as they grew up so they wouldn't make bad decisions in their lives that would result in them being sentenced to prison. Vandrick considered himself pretty strong, both mentally and physically, but incarceration was a process of adjustment he was not built to bear indefinitely.

CHAPTER SIX

Co-Parenting Through Concrete

Through trials and errors we were learning to raise our son together from behind prison walls. It took various adjustments, but we had gotten better at it. Most families had the luxury of discussing their days over the dinner table or while sitting on the living room sofa every night. With prison life, you have the luxury of catching up after there's been a long span of lock-downs with no phone calls and no visits, waiting on a letter in the mail to make sure everybody was doing okay. You go through issues, cares and concerns at home, writing them down and dropping them in the mail then waiting on the call or the mail response to an issue that needed an urgent response. Of course, when the call finally came, the standard question of, "Are you okay?" was asked even before you attempted to try to catch them up. Anything could have happened between the last contact and the current one. The relief for you was that you got a call from that recognizable number which indicated the caller was still alive and breathing behind those concrete walls.

Some couples have relationships throughout their loved one's time in prison without any children being involved. It's just the two people, inmate and non-inmate. That would be an ideal situation. The only people that were of major concern are the two of them. But that was a rare occurrence as I had started to witness. Most inmates had children, even if they were not their biological children. Those relationships took far more finesse to navigate through when your help was miles, hours and extended time away. Since I was the type of woman that would not terminate the relationships between my children and their fathers, I implemented certain routines into our lives. I don't just use their fathers for the sole purpose of the disciplinarian. I was just as much a parent as their fathers and since the children lived primarily with me, discipline was something I enforced in my household. I did share this information with their fathers and explained that when any discipline was warranted, I was not waiting on delayed mail or phone calls to maintain stability inside my home.

With the introduction of Vandrick's prison stay into our co-parenting arrangements, I did not want to leave the negative behavior for his dad to deal with while I provide the nurturing. I learned all of it had to be equally shared between the two of us. So the moments to celebrate a job well done and the moments to lecture over defiance or grades had to be equally distributed among both of us. His dad would miss many of the fun moments of birthday parties, graduations, special events, vacations and other such things. However, I took pictures of as much of it as I could as I tried to capture those moments for his father. Good report cards I never failed to mention and even sent a photocopy. We would write, send cards or letters and if we forgot we would receive a friendly reminder during a phone call or letter in the mail asking, "Hey, can you send me those pictures?" Then I would do my best to make sure to send them on my next trip to the Post Office.

Our son was getting older and a display of a united front was becoming more necessary. No matter whether both parents were in the home or not, a child would attempt to try and pit one parent against the other and I was not going to allow myself to be blind-sided by that tactic. I knew how to get ahead of it. Therefore, his father and I needed to make sure our communication with each other was not fragile. Not having a man in the home made it more important that our son realize it too, no matter what. There was an all-around support system to keep him on the right path. One of the things that had been introduced into the prison through corrupt prison employees, officers and free-staff alike, was making sure inmates had the ability to communicate through the illegal access to cell phones. Whoever was trying to make some extra money on the side of their already plentiful prison payroll check was into this underground business. They found supply and a hefty demand existed in the prisons.

Cell phones in prison were expensive from what I'd been told; sometimes double or triple the price of an actual legal phone on the street. If an inmate had access to that type of money, they were able to have that luxury, no questions asked, just provide the funds and the product was there. The prisons were filled with them and inmates were now able to stay in contact with family, friends and even enemies as if they were out on the streets. Those lock-downs and constant loss of communication were becoming far less of a burden for prisoners. I am sure the surge of cell phones in the prison was causing the postal services and contracted telephone companies to take major financial hits. Then a massive raid to confiscate the cell phones was implemented and inmates had to return to their traditional means of staying in contact.

I knew without a doubt those same corrupt prison employees also made sure they stayed in business because they would unexpectedly drop a dime on which inmates had cell phones and have their cells raided to make sure they could keep their extra side-hustle money rolling in. There were parts of society that didn't have any connection to prisons or its inmates and would assume the incarcerated criminals were the only ones committing crimes. This was definitely not the case. Some of the officers and staff tasked with the responsibility of maintaining their integrity and discipline were no different than the inmates they were responsible for monitoring. From time to time I heard from Vandrick via cellphone, but for the most part he would write and utilize the collect calling system. I always did my best to keep money on my phone so there would be no excuse for missed phone calls. Our son needed to know that his dad was on board and doing his best to be a contributing participant into his life as best he could and through whatever means that were available to him.

One evening Vandrick called via collect call and I was just in the middle of fussing at our son about getting his classwork and homework completed. It was important he got good grades and stayed on top of his assignments because once you start missing assignments the harder it becomes to keep up. I think right at the point I got really angry the call came in and I was so frustrated that once I accepted the call, I just blurted out, “I’m glad you called! Your son needs you because I am going to lose my cool with him!”

“Where is he and what did he do?” he replied calmly.

“He’s in my room crying and he doesn’t feel he needs to do his school work,” I told him.

Vandrick quickly took over from there with a simple, “Give him the phone.”

I gave him the phone and walked into the living room where I needed a timeout more than anyone else in the house. I was counseling myself as a barrage of questions and thoughts ran through my mind: “*Does our son not understand the importance of an education? Does he not understand the sacrifice that is being made to make sure that he has access to the best education that I can provide for him and his sister? What is he thinking? Does this boy not get it? There are children that wish they could be in the position he is in? What do I need to do to get him to always want to give his best effort and not settle for less?”* While I was in the living room not knowing if my own questions were further fueling me, or if they were becoming points to rationalize my son's perspective from a child's point of view.

I could hear their phone call coming to an end. Our son was slowly making his way towards me with tears still on his face. He slowly dried them with his hands as he looked at me and said, "My dad wants to speak to you."

I took the phone from him and I was being scolded by the most unlikely of sources, "Hey, you need to calm down. He gets it, but he can't process it with you yelling at him. He told me he will get his work done and you can check it in a little while."

I became less tense as those words were coming across the phone line and directly into my ears. I told him I would follow through on my end and I appreciated his call because timing was everything. I don't know if he had called to tell us anything in particular because the entire talk time was dedicated to this active parenting moment. But, if he had needed or wanted anything else, it would have to wait until the next available opportunity as this call was over. When they say a 15-minute call, they meant it because the phone was going to disconnect whether you hung up or not. It's just a rule you never got used to. After the call ended and our son finished his assignments for school, he did say to me, "Mom, I apologize for making you angry." I don't know if his dad coached him to apologize or if it was of his own free-will, but I appreciated it and gave him a hug as if he'd initiated it all on his own. Our son did have a soft and sincere side that he could push out at the most convenient of times, so this was likely one of his naturally charming moments.

Other moments of co-parenting happened when things would build up on a daily basis and we'd have to discuss them in the visiting room. I was getting better at visiting, even bringing one of his daughters with me when his wife couldn't visit. It was a normal visit and she had been the most frequent of all his children to go and see him, so she knew the process better than I did. It was nice that we brought her along. She was a mild-mannered, easy to maintain child; a very mature and kindhearted young girl. It did help with more than one child going to visit because they could interact with each other. That left time for Vandrick and me to talk and I was not just sitting and looking around the room for new things to discover. The kids played together, everyone ate, then there was playtime for them with their dad, not to mention the stern lecturing time also available for him to get some key points across to his kids. After all that was accomplished, there was little time remaining. It would be pretty close for the visit to be over if we timed it all right.

In preparation for this visit I did as I had normally done. I brought extra clothing, blankets and pillows, games and snacks to keep them occupied for the car ride. Our son stayed up a little bit longer this time, I guess to show his little sister that he could stay awake with her, but that was short-lived when they both fell asleep from the long day and exhausting trip. I received my routine call to tell me, "Thank you" and he was even more appreciative that I had gone by and picked his daughter up and would drop her off at home when we arrived in the Bay Area. I was always letting him know he was welcome. After arriving at his daughter's home, I sat and talked for a little while with her mom. We got along well and had similar desires for our children and both of us believed in making sure our children maintained discipline and respect. Plus, the extra time talking was giving the kids more time to play, or pretty much drive each other crazy at this point. Then my son and I gathered ourselves and made the remainder of the drive home to get cleaned up and some much-needed rest. Even if my son didn't need it, I surely did. All that driving back and forth took its toll on me and I always needed some rest, even if it was a short nap after visiting. It never failed.

I noticed during the visit that Vandrick had some things on his mind, but I chose not to pry. He would tell me what I needed to know in his own timing, but I noticed something when we were parting ways. The only thing I was certain of at this point was that he was maturing and going through some transitions in his life even if unspoken. I haven't asked anything about his case in a while so maybe that's it; maybe he's going through some transitions with trying to get his appeal processes moving in the right direction. He didn't look mad or angry. It just looked as if he had a lot on his mind. But who could blame a man that was locked in a box and trying to find his way out?

During this time he had a cellmate that he had good fellowship with. I knew this was the case because he told me what he knew about him and he would never introduce or speak to me about anyone that he didn't consider close to him. Also, this brother knew lots about me and I knew lots about him, so much so that if for any reason I could not locate Vandrick, or Vandrick needed to contact me, we were allowed to speak to each other. Unfortunately, the dark side of prison that people avoid discussing was what made protocols like this exist. If something happened to a loved one and you're not the first line of contact, it might take a long time for information to reach outside those walls. Therefore, making sure I had a source for information remained vital for us. There might be a time when I don't know that Vandrick ended up in the "hole," more formally known as Administrative Segregation (Ad-Seg). Therefore, to avoid an unnecessary long drive because those visits were by appointment only, whoever was trusted to contact us would let me know not to make the wasted trip. Things like that were important and those courtesies were appreciated.

Vandrick also trusted this brother because he was becoming more committed in practicing the religion of Islam. Vandrick had followed the practices of Islam, even out on the street, but at that time he was more concentrated on the Nation of Islam. They were more visible in Oakland and we would see them walking or driving down the streets; the brothers handsomely dressed in suits and bow ties; the brothers calling out to those listening, "Final Call" as they held a newspaper in their hands. I was born and raised in the Christian faith, but I did pay attention to the structure and discipline displayed in the Muslim community. It was good for me to see because it provided a different image to observe versus the young black brothers that were losing their lives to turf wars and drug deals gone badly. So I assumed that was also something that intrigued Vandrick's also and although he and his family were Christian, he started to study the teachings of Islam. And now with a Muslim cellmate and all the lock-down time they had, they were able to really spend that time learning the Deen of Islam. However, Vandrick's interests were transitioning from the teachings of the Nation of Islam he had seen on the streets and were becoming more focused on the ways of the Sunna.

He was becoming so captured by the religion that he would start to learn the meanings of the names based on the characteristics and qualities of an individual. He would soon carry the name of Hamza, "the Lion"; it is fitting for him because he's born in August, a Leo. And has the demeanor of a lion when challenged. He would also give his children Muslim names as well. Our son would find out that his name was Malik, which was ironic for me to hear for the first time because that was his intended middle name that I found when I was pregnant with him from a book of names. The name Malik, means "King."

I always wanted him to recognize himself as such and hold himself to that high esteem, but his name was abruptly changed during my labor and delivery by my sister who decided that she in some way should be included in deciding my child's name, even though she had already had a son of her own that I did not participate in naming. Let's just say I was not pleased at the change, but once it was recorded, I wasn't up to fussing about it. I took the same position in not changing our son's last name since his dad wasn't present at his birth. What's done is done. The irony of it all is, Vandrick and I both were unconsciously in agreement that Malik was definitely the name. It was always supposed to be a part of our son's name and Vandrick knew nothing about it being our son's original middle name until it was mentioned during visiting. Our son liked the name as well and I explained his name to him and let him know I felt Malik was definitely destined to be his name.

CHAPTER SEVEN

Don't Take No Wooden Nickels

Well, before long, the struggles of single parenthood would soon force me to make a tough, but sensible decision. With being a student trying to progress my career, financial responsibilities and obligations intermingled with all the other challenges I was facing, I had a new, yet old friend, come back into my life. I believe at that time he was repositioned into my life to show me an avenue I would have never in a million years ever taken seriously if he had not reappeared during that time. So, on November 19, 2008, I enlisted into the United States Army. It was not a sudden decision. I had pondered the possibilities numerous times, even trying to find ways to talk myself out of this process. However, the more I turned away from it the more it became my only option to see myself complete at least a small portion of the mission I had set out to accomplish.

I realized I could become a nurse in the military if I completed my courses successfully. That would certify me as a licensed nurse and I could achieve the other degrees soon afterwards. There was a bonus guaranteed for the career path I had chosen and I absolutely could use the money. I felt that with another source paying for my education—since the books and other things were all coming out of my household budget—it made financial sense. Therefore, while I worked and gained the knowledge they were offering, I could simultaneously receive a constant paycheck.

The other thing that I took into consideration became an incentive for me was my education would be attainable with minimal distractions. I could continuously attend school until I completed all courses. I would have to enter into a guaranteed, 8-year contract that would be broken down into 2 years of active duty and 6 years in a reserve unit. This was an extremely tough decision when you're only 30-years old and have 2 children that depend on you on a daily basis. I didn't know if I could do this at all because it was not just them depending on me, I now depended on them as well. They motivated me to struggle through some very tough challenges. I wake up daily, giving thanks that God chose me to mother them, but seeking His guidance on what was best. This decision would come with many tears, lots of fears, but confidence that if I could get through some of the things of my past I could get through this as well. I wasn't sure, but it was what I had definitely started telling myself; you can do anything you put your mind and heart into. My self-talk had to be louder than my nerves or this would likely never happen.

Luckily, I had the support from my former high school sweetheart who had set aside our past issues and come back into my life at the level he could participate on; who had already served 10 years in the military and had recently come to visit the kids and I during his 2nd deployment to Iraq. But I also had support from my mother who at first thought I was joking, but soon realized I was serious. As a matter of fact, I think everyone thought I was kidding, crazy, or something close to that. I was the woman who showed up regularly to my office job in 4-inch heels or higher, my hair and nails done. I was very spontaneous as many would agree, but the military, hmmmm… No one ever saw that coming, not even me!

My dad was on board, too. He was former military as well, although a different branch. He was in the Navy when I was a baby, but I wanted to stay on the ground and joined the Army. His favorite saying, no matter what I was doing during this time of preparation was crystal clear, "Don't take no wooden nickels." I came to understand that this would be useful information to members of the military, warning them to not let the military swindle you out of what you have earned or agreed to. That was a good warning in basic training because the first question asked was, “How many of you were told once you got here you would be able to change your MOS (Military Occupational Specialty)?” At least 50% of the newly enlisted soldiers raised their hands.

My grandfather on my father’s side gave me his seal of approval over the phone and he told me I was about to walk the path that he first paved. He was an Army man. My sister and brother-In-law thought it was honorable and they would become the two I relied on and who I would have to soon sign the rights of my children over to. This was rough, but a necessary requirement in my enlistment process. Everything was legal and by the letter of the law. I had to have reliable members of my family to leave my children with. My children’s fathers, well let’s break them down since there are two. My daughter’s father was partially okay with it, but for the most part he was against it. He felt I needed to stay home and raise my children despite what he knew was transpiring around me and the burden of the load that I was carrying. I understood his position, but if it was up to just him to make this decision and he was my only line of support, I can honestly say I would not have been able to join the military. Then there was my son's father. From the moment I told him, I don't think I'll ever get over his response.

"What in the hell did you just say? Qiana?" he almost yelled. "I can always expect you to do something that no one else does."

I'm not sure if he agreed or disagreed, but it was a response nonetheless. His silence as he was in deep thought spoke something more meaningful, but he wasn't forthcoming and I wasn't willing to force it out of him, so I continued forward with my mission.

Everything was final and everyone who needed to know now knew. The rest was up to me. I was joining the military and that was it. I tested high enough to get the MOS that I wanted. I had to fight a bit for the benefit of my children and myself to attend my schools consecutively because I would have to leave the state to complete this mission. I got my request with a slight change in my contract, but it was worth it to not have to disrupt my children's lives as much as possible. I signed every page of the contract and thanks to the help of my friend, my former high school sweetheart, my contract was signed according to the terms that I was comfortable with agreeing to. So far, "I haven't taken any wooden nickels. Thanks Dad, I am listening."

As I went through this process, I became a nervous wreck which is putting it lightly; an emotional basket case was an understatement. I didn't leave for almost eight months after swearing in and signing my contract. I went to my reserve unit when they had drills to acclimate myself to the environment of the military. I can appreciate this exercise because I was becoming more confident in the things I was able to see. I learned time management, the different ranks, different marching movements and how to properly perform them through trial and error. I even learned how to properly fall in and out of formation. I was, well kind of, transitioning from an office diva into a squared away soldier. I knew more now than that first day I showed up in my thick sweat jacket, jeans and what felt like my first pair of tennis shoes ever.

It was funny to me that soon after joining they would give me my uniforms so that I could wear them at the unit. However, I often had to remind people that I am pre-basic and I don't know much although I am getting used to military etiquettes rapidly, especially after getting a stern talking to by an E-7, who saw me salute our

Commander at the time, Captain Ang, inside a building as he came through cheerfully, greeting every soldier he laid eyes on with a pleasant smile and "Good morning." I was able to take my son to the unit a few times for family day events given prior to my departure. The Christmas event was the one I recall most. We woke up early to arrive at the unit inside Moffett Field where they had a day full of plans to accommodate the children although the soldiers had prior work to do. The kids would be occupied until we would have a lunch time holiday meal provided by the Officer's side of the unit. It was definitely a nice experience and a great way for a newly enlisted soldier to meet the unit.

I was slowly trying to take it all in, but with thoughts to follow. Soon, the day of my departure seemed to be coming far faster than I ever wanted it to come. No one and nothing could prepare you for what happens after you've signed all the guardianship, power of attorney, wills, opening bank accounts and the other things necessary to transition your children over to other family members. There is no documentation that exists that stops the river of tears and emotions as you are moving your kids into what will become their new home for the next two years. There is no manual, guide, or regulation that prepares you to drop your now four-year old baby girl off to her first day of kindergarten and to stand there, looking at her start her day and realize it is the hardest day of your life. You and she know this wave goodbye and that hug you got was the last one for a very long time as you pray that she will understand.

I dropped off my son, now 12-years old, at school first that day. I always mentally told myself he was going to be my easiest child to handle this because he was my oldest and would be able to understand the necessity of the sacrifice. I was so confident in my belief. He didn't cry when I dropped him off at school. I don't know if it was because he was in front of his middle school friends and it's not cool to cry, or because he was able to handle the adjustment. After all, he was going to be housed with his other boy cousins so this might be a cool experience for him. He would develop a brotherly bond with them. All these things I was telling myself to make sense of it all.

I had been told by my recruiter I would leave early the next day after the first day of school, but that didn't happen. There had been a mix up on the departure dates. He had it wrong by one day, so I showed up at the airport on the wrong day with my mom. She was my transportation to the airport and she quickly told me to gather my bags and get out of her car because she didn't want to see me cry. She knew it would only make her lose her cool as well and we would both be no good. She dropped me off at the curb and kept right on back to our old house in the old neighborhood. It was extremely close to the Oakland International Airport which is where I was scheduled to depart. I had to call my mom to come back to the airport to get me. I was laughing as she thought initially, I was calling for her to be my co-conspirator in hitting the highway because I was deciding to go AWOL. I'm sure if I'd said it, I could've convinced her to do it. I don't know for how long, but she would have been ready to roll if I couldn't handle it. I had to get her to understand that this drop off would happen again, but on the next day. I did not make any plans to put my children through the process of saying goodbye for a second time so I remained at the house in Oakland and didn't make any calls to them. This allowed me to stay with my mother one last time then we did a repeat the very next morning.

Yes, she put me out again at the curb just as she did the morning before. She gave me a brisk and dry hug and said, "Goodbye" as she drove off into the direction of our old house. My recruiter and another female soldier heading to Fort Jackson were there, waiting for me. The female soldier would board the same flight as me. Our recruiter walked into the airport with us as far as they would allow him to go and he bid us farewell as he reminded us both to call if we needed him for anything. But he said he would be following our flight plans closely until we made it safely into South Carolina. Thankfully, I wasn't traveling alone, but the pain in my heart sure had me feeling like it.

This was the first flight ever where I didn't immediately fall asleep. I had a window seat, my favorite place to sit because I wasn't making several trips up and down the aisle and I didn't want anyone constantly climbing over me. I gazed out the window, meditating and praying that I was making the right decision; I asked God to cover and protect my children in my absence, and that He would help them to withstand these sacrifices that we all would endure; and to give me the strength to complete this mission. Tears fell down my cheeks as I silently prayed and eventually, I found myself falling asleep.

CHAPTER EIGHT
Sound the ALARMS

Before I left, I made sure I talked to Vandrick on the phone. I didn't have much time during the process to visit him like I had before and things were rapidly transitioning around him as well. Salinas Valley had become very hit or miss with visiting and they were constantly on some type of extended lock-down. So, we made it a point to talk on the phone as much as possible. He was aware that writing had become scarce on my end because I was moving from my home and moving the kids in with my sister and brother-in-law. Lots of moving pieces were taking place and it would seem that it was all taking a long time, but for me, the person who was doing the moving, things were happening fast as I was busy making sure every detail was taken care of such as: furniture had to be placed in storage, all bills finalized and closed out and making sure pending bills that would remain were identified because I was no longer making sure they got paid. I had to transfer permission and access to my bank accounts over to my sister. These were things the average person didn't have to think about but was now becoming my reality.

Now the routines of our co-parenting were being integrated into this process. When talking on the phone we were not just shooting the breeze, we were making plans for our son. I was making sure all documentation was in order for my sister to release my son to Vandrick's wife to take him to visit him in prison. These documents had to be notarized. I let him know that during basic training, calling the kids would be extremely limited for me so I would need him to pick up the slack on his end whenever possible. I would have my cell phone, but it would be locked up most of the time except for about 30 minutes or maybe less on Sundays. I knew I couldn't mother two children, 30 minutes a week. So we were trying to work out his plans to be able to call during homework hours to make sure our son was still doing what he knew he was supposed to do. If he had 20 minutes of required reading time then his 15-minute call would be the time for him to complete that.

We were working it all out. If our son needed to see Vandrick urgently, then he could do that. I felt he needed to know one of us was always accessible to him. I believe Vandrick was now more aware of the decision I had made in joining the military and the effect it would have not only on our son, but also on me. It was becoming more apparent that my access to the kids would be a direct reflection of the life he was currently living. I know this was an accurate perception on my part because during this time he promised me, "Qiana, I am going to ride this time out with you. You will always receive mail from me." It was refreshing to hear whether it was true or not. The encouraging words of support from a person who could understand the lifestyle of confinement was refreshing and calmed some of my fears. We had our strategic planning all mapped out for our son, and Vandrick was fully on board. Even though we never really dated, we were always each other's friend and that mattered so much more than establishing a romantic relationship that neither of us had been ready for.

Unfortunately, even with the best laid plans, something could fail. But at least we put the work in to try to make things as smooth as possible. However, I hadn't taken into consideration that my family was not on board with our son visiting his dad in prison. No matter how many times he had gone before, and some visits without my presence, it was still an opportunity for them to implement their own ideas despite my request and all the preparations Vandrick and I had made to make sure it would be as easy and as burdensome free as possible for them. I should have paid closer attention to those subliminal messages and sidebar comments indicating their objections long before I had ever left.

The transition for my daughter did not proceed without glitches, either. After arriving at basic training safely, I spent a couple days in the reception center making sure we were set to move over to the full training barracks. I was able to call as I first landed and let the family know I made it safely and for those couple of days we did keep our phones with us, although locked up in lockers the majority of the time. However, nighttime was likely the time, if you weren't passed out, that you could text and let everyone know you were still alive and well. I sent a few texts, but transmission in the reception barracks was weak at best, so as I was sending the texts out, I was not sure how long it would take for them to be received. Most times I fell asleep waiting.

The random wakeups in the middle of the night and morning for mass punishments didn't help at all. I would not be able to make an actual call until I was allocated time on the CTA of the reception center. So, I dialed my mom's phone number, finding she was out with my sister and son at a school sporting event. I was happy to hear that, but was a bit puzzled when they put my son on the phone and after talking to him for a few minutes I asked to speak to my daughter and my son hesitantly informed me she was not with them. He said it in a tone that alarmed me because it sounded as if he had been coached not to tell me. However, my son would not honor that request because he knew that he was not supposed to hide anything from me, whether good or bad. So he tells me that since I left, his sister had not been with them for about three or four days at this point. I didn't want to have him in the midst of grown folks' stuff so I asked to speak to my mother to find out what the hell was going on? Where was my daughter? And why hasn't anyone gotten information to me? Then my mother tells me her dad had un-enrolled her from the school she was in and enrolled her in a school in the city where he lived. This would have been completely acceptable if he hadn't had me served to appear in Family Court almost a year prior and through mediation, we had to determine all our daughter's needs spelled out in documentation. Let's just say this was the first time that anyone in the military would see me go bat-shit crazy and have to come to my aid with urgency!

My mother is on the phone, trying to tell me they have involved the police and provided them with the documentation that I had left, including the packet with the legal papers from the Family Court Mediation Department. She is patiently trying to convey to me that they are working to get my daughter back from him peacefully, but unfortunately had to involve my recruiter and his command as well. However, my recruiter was trying to avoid reaching out to me via an emergency Red Cross notification on the other side of the United States. After over a year of knowing me, he rightfully suspected I was going to hand over my uniforms to whoever needed them and immediately fly out to go get my baby without a single care in the world about the consequences for every last one of my actions.

The time allotted for phone calls was over and we were moving into our next task of the day which I believe consisted of preparing us to move to our unit. I was furious at this point! My daughter, in my opinion, was by all rights missing and I was about to explode. I prepared to get to my unit with thoughts of my daughter's possible whereabouts racing through my head. I was trying to calm my blood pressure down with my mother's last words, trying to convince me they were on top of things. It was not working, but I was trying. I needed something or someone to coach me off the mental ledge of pure insanity. Then I thought, *"How long had her father been plotting this act?" "Was he crazy enough to try to run away with her?" "Was this out of anger at me for leaving, or better yet, because I decided that I didn't want a relationship with him?"* I wondered if every other disagreement and act he had attempted to gain my attention had been fueled by my not getting back with him. I guess the random craziness we went through in our relationship didn't give him enough of a hint to the side of the psychosis I will tap into when you detonate it by crossing that explosive line of messing with my children! I see RED!

I was on the military bus and we were being transported. We were told to put our heads down to prevent us from seeing where we are being transported to. I am willing to obey the command even though I am not in the most stable of places mentally or emotionally. Nothing any drill sergeant or soldier had said to me since I made that phone call had registered with me. I was just following people who were dressed like me to wherever they were headed so I could talk to someone with some authority. I wanted to specifically give them a heads-up that I was leaving and the only way they could stop me was by lethal force.

We made it to our unit barracks and after an introduction and a little hazing we attended our first meeting in the bleachers underneath a tree for the shade. I had noticed the incredible difference in temperatures of September in California to the humid heat of South Carolina. I was soon to be introduced to whoever I had been waiting to see; people with distinguished authority over me; our 1SG (First Sergeant) and Commander. During their introductions to the unit such as rules and regulations and things of that nature, some information was being put out that might be resourceful, but I had no mental capacity to grasp anything. They finally offered the opportunity to see us individually regarding any concerns basic training soldiers might have. To establish proper military etiquette, they even reminded us that if we meet with the Commander to stand at the position of attention; if it's with the First Sergeant then it is parade rest. Yeah, yeah, I got it was my thought while my mind was focused on needing to be seen by someone immediately.

We were lined up and I think I was about third in line. I walked towards the First Sergeant but instead, I don't recall how, found myself standing in front of the Commander. I was trying to formulate the right words to say and use proper etiquette, but all of that flew straight out the window. I broke down in uncontrollable tears as I tried to let him know that my last call home was with my family, trying to get my daughter back from her father and she had been gone since I got on the flight and left California. I am sure it didn't come out as fluently or coherently as I'd hoped because it was all a jumbled mess and there was no point in me regaining my composure at this point. I had been holding it all in far too long. I am also certain I was not at the position of attention, but I think he got the gist of things pretty quickly.

The Commander didn't even expect me to remain professional at this point. He knew sensitivity and urgency overrode diplomacy in this meeting. I needed help with getting my daughter back or she wasn't going to be the only person missing. The Commander stopped everything. No one else was allowed to speak with him. I believe the rest of the line was shifted and whoever else had an issue was now shifted over to the 1SG. The Commander took all of my information and copies of my documentation that I had with me. Even if my recruiter didn't want to make a fuss from his end, we were officially making a fuss from my end back to California. My Drill Sergeants were meeting me from day one on a very personal level. They learned my last name almost immediately. I was allowed to make calls and try to get some answers. I needed reassurance that my daughter was being returned. I received confirmation that my daughter was being un-enrolled in that school near her father and placed back into the school she was originally in, down the street from her aunt and uncle. She was guaranteed she would be back with them by that Sunday. It was Thursday when I called and for sure I would be calling back on Sunday to make sure. If she had not been returned, I was going to get on a red-eye flight back to California with no questions asked. I would like to believe that my daughter's father did not like all the attention that he'd gained in a matter of days and whatever his plot was it had been foiled swiftly. I stayed where I was, allowed the resources I had available to work for me. Even though I knew this was not my first plan, it all came together better and more peaceful than the one I was ready to initiate.

Well, I was adjusting to the military lifestyle and had built a routine once I got through the storm. I started to realize it wasn't much different than prison life with a lot of dictating, rules and regulations. It was slightly better as there were some good things about it, but for the most part the structures were extremely similar. In the evenings before the last formation of the night which included our mail call, we would have a bit of free time. This was a time where we could get to know our "battle buddies" better, especially the ones from different platoons. We bunked eight females to a room in the newer barracks. This was much better than some of the older barracks where there were 60 females in one common area known as a bay. We were spread out down two long hallways separated by a set of stairs to get down to where our platoon classrooms were located and the Drill Sergeants as they pulled night-duty. After doing sets of push-ups, sit ups and squeezing in a shower, this was also the time that I made sure I would dedicate to writing to family and friends so I could remain in touch.

During final formation I would gratefully receive mail pretty consistently, thanks to my co-worker that always promised me he would support me out here on this mission and my friend I grew up next door to in the old neighborhood who was also incarcerated at this time. Vandrick did as well. Mail from my kids was delayed periodically, but when it arrived it was always predominantly school projects to show me their grades, artwork that my daughter had drawn whether at school or in her free time at home, and even a few handmade cards that said, "I love you mom" mixed in with bills and mail that had piled up in my post office box. No matter what it was I made time to look through all of it and put them where my personal things were stored in my locker. That mail helped to keep me connected because the internet was no longer an option I could rely on. When I was in the field and there would be no mail delivered, I kept a letter or two with me. When it was late at night and I didn't feel sleepy I would take them out and with minimal light I would read them. It helped me to feel closer to home. Some soldiers had snacks in their cargo pockets, but I had at least one letter from someone back home. I had to substitute for their absence any way I could.

Vandrick and I stayed in contact and he even wrote to notify me that he had been transferred. He was no longer at Salinas Valley State Prison and was now at Folsom State Prison. He also let me know he had gotten divorced. Well honestly, I had known the divorce was in process before I left but I didn't take it seriously or believe it because they seemed like they had a pretty good thing going. I thought they would reconcile any differences they had, but as with any relationship you never knew what was going on underneath the surfaces. So I apparently was proven wrong as I read the sentences of his letter. He was getting married again to another lady. She had no children by him which in itself was surprising because he knew her prior to going to jail. They went to the same high school whenever he decided to attend. Those were really the years that he was in the streets, hanging out with the other kids, not going to classes, selling drugs and less than concerned about his education.

Vandrick didn't graduate from high school with his graduating class, proving he hadn't spent much time in a classroom. He didn't mention much else about her and his description of her was vague, almost non-existent. We spoke about many things over the years and this was not one of them. I wasn't sure where his head was at when he made the decision because it seemed a bit sudden after his first marriage to be talking about a second marriage. However, I would support him as a friend should. He wanted me to know about the changes taking place in his life and I had no reason to object to them. What I did was write him back, letting him know I received the letter and I wished him the best in this marriage. It was also necessary for me to write to him because the name I had placed on all of the documentation for our son to visit him was his first wife's name. I would now have to try to convince my sister to change the name of the documents to his fiancé's name, if our son wanted to visit him.

We normally would discuss all things like this either in person or over the phone, but under the current circumstances I would have to rely on the mail. I did write and try to give him an approximate time frame when I figured my cell phone would be given to me, but it was a complete guess based on prior Sunday. There were no guarantees as I was learning that plans could change in a matter of minutes and I needed to be ready for those adjustments. I would have to just hope that he received the letters and see what form of communication he would use for his response. My military lifestyle was definitely mirroring that of any prisoner at this point. I was distant from my family and relying on the military's long-standing institutions for my sources of survival versus what I had become accustomed to: being a single parent in my home life where I was the glue in which all things, including my children, were held together. I was beginning to understand far more deeply the type of encouragement Vandrick needed from the beginning of his confinement and even now. We're both realizing that this encouragement would have to be an internal process, especially on the days when the comfort of familiar things was so far out of our reach.

I would endure nine weeks of obstacles, new perspectives, learning new things not only externally, but as well internally. I was becoming stronger mentally and physically and truly understanding the meaning of fortitude. I suffered an injury very early on, but my passion to see my children superseded any pain I would ever experience. I had to put it on the back burner in order to complete my training on time and make it back home within that two-year plan as I'd promised my children. I would not let them down. I was not the only one counting down calendar days; they were going through this with me. They needed their mom to get through this mission successfully and make it back to them to be there to love and motivate them up close and personal. They were what drove me to just keep moving, preferably in a forward direction.

High and lows, great people, lasting friendships and motivating leadership helped to square us away. We rushed through the cycle as our leaders were exhausted from working two basic training cycles back-to-back. We were reminded regularly as we powered through the phases of Red, White and Blue. Discipline became more lenient as we learned to follow the rules within the confines of the military's design. DS Johnson, DS Wideman and our third DS we lost due to poor decisions he made during our cycle, were motivators and instructors of experience that no solid lesson plan could ever teach. The very memorable, "Wholly Sheeit!" called out by DS Wideman, as only he could, if one of us messed up; or DS Bear, our replacement Drill Sergeant an actual foot tall teddy bear wearing the military uniform, who would come flying in through the opening of a closed door that we would have to holler "At Ease!" and stand at parade rest because DS Bear had as much authority as our actual Drill Sergeants. Fun was an option. We were rarely referred to by our given names, but more memorable nicknames that spoke more honestly to who we were, individually. Or, many times the 4th Platoon were left out on the field somewhere while the other three platoons got packed and were ready to go. Jokes and downtime allowed for more bonding outside of work hours to develop better understandings of individual personalities.

I was not the only single mother enlisted and driven to succeed for the sake of our children, surviving through the changes at home. We mothers had a unique understanding and a purpose that couldn't be explained why we were pressed to complete this mission. It was what we were carrying in our hearts that was greater than ourselves. Soldiers, parents and non-parents alike leaned on each other, literally and figuratively, to remain dedicated to getting the rewards of our cycles on graduation day in quickly approaching November. These pictures and memories would play out for us all on graduation day—the good times and the bad times, our insecurities and our new found confidences on display for family and friends to get a glimpse of what we had become, soldiers. We were proud, United States Army Soldiers. We have declared to take on the task of the next phase in our careers, dedicated and committed to "never leave a fallen soldier behind." The bonds of Battle Buddies were created within a short nine weeks of a Basic Training cycle.

Vandrick is married now. He sent me a picture of their big day. It was nice to see, but the message on the back of the picture was a bit mind-boggling. But as I have done many times before, I ignored his slick comments, laughed and kept right on point, knowing he couldn't possibly have meant that message on the back for me. He had a way with words which were always attention-getters even if they sometimes threw me off a bit. I didn't question it because he just might tell me what he really meant then I would be held responsible for what I knew. Anytime he sends a comment like this I intentionally let it go over my head. Yes, we had a past of hooking up out of convenience, and he always felt that no matter what, he still had rights to me, even more so after having a child with him. But he was married now so if nothing else I figured he'd made a sound decision where he wanted his life to go and who he wanted it to go with. As I said to him, "I wish you well." I don't know what made him decide on something as final as marriage, but I hope he was sincere in deciding on his life partner because at this point the only other thing I could say was, "Congratulations!" I would find the time to write him back during our personal downtime and make sure he received that acknowledgement from me. I know he would say the same if it were me.

Vandrick always told me whether he agreed with me or not on my decisions, he always supported me; and even if I failed, he did his best not to bombard me with a bunch of, "I told you so!" He allowed me to talk through my decisions, both good and bad, and he was there to help me find my way back to reality. I can't do anything more than respect him for that. It is one of the things I admire about him as my friend. As a man, and now a dedicated Muslim, it is plain to see that underneath that rough exterior from losing his way in those streets, he was still raised with some sense in his big ol' head.

I will miss the upcoming 5th birthday of my daughter in November, but I did the next best thing. I paid for my mom to travel and bring both my children out to see me for Family Day and Graduation Day scheduled for the early part of November. They traveled with one layover in Chicago where my mother unfortunately reinjured an old wound coming down an escalator. But nothing was going to stop her from making it to South Carolina to see her daughter and to bring my children. My kids did their best to help their grandmother as much as they could when she could no longer bear any weight on her now swollen ankle. She had to be carted in a wheelchair to their next flight and finally from that flight to her waiting rental car. She hobbled and pressed on until she got into the car and made it safely to the base.

She followed all the directions to get on post with my two children in tow, where I was patiently waiting for their arrival on the morning of Graduation Day. I hadn't seen my children's faces and they hadn't seen mine since our departure the morning of their first day of school at the end of August. They easily found me, based on my height. It never fails because I am always on the front row of anything military—shortest in the front. After the graduation ceremony came to an end, I stood on the flat ground floor of the auditorium because I had a limp I often tried to ignore and mask. They would be coming from the bleachers. The first to leap into my arms was this flying princess in a big puffy dress, my daughter. I was glad I could catch as she sprang into the air and landed in my arms, tightly hugging my neck. With a huskier junior football player silhouette, walking toward me swiftly and tears streaming down his face he could not control was my son. He was weeping and grabbing me so tightly I had to widen my stance to hold all of us up. And in that moment, I realized with one child smiling as she saw me and one child crying, my thoughts had been completely wrong. My daughter would be okay because she had my mom, my sister and her dad to comfort her in my absence. She would adapt to the changes in our lives far more easily. It was my son who did not do well. He was so accustomed to my presence, my involvement, our connection. And now he was no longer seeing his dad as well due to my family's lack of understanding that my son already had a relationship with his father and needed to continue to visit him, this was not an easy transition for him at all. “Wholly Sheeit!” I messed up!

We spent the remainder of the day together. I bound up her injury and borrowed crutches from our stash at the unit for my mom, promising to return them later in the evening, but she would need them throughout the day. We toured the barracks and I showed them a few places on post. My Commander remembered the initial scare I’d had from our first encounter and now he would get to meet the little girl that caused all of us to almost have mild to severe heart attacks. My mom, the kids and I ate together and she drove us back to their hotel room. We were all so tired plus; they were all jet-lagged. So, we slept until the alarm went off and I realized that I had to get in full uniform and back on post before final formation.

The kids got to see formation. Boy, did they get an eye and ear full. Of course, with the opportunity at a little freedom a soldier or two would show up late. So they also got to see what getting “smoked” looked like. We received full, mass punishment to remind us that we were still soldiers and held to a higher standard. As my son looked on, at that moment, he decided he was not joining the military from the screaming taking place. I had become numb to it, so it sounded like talking to my ears. He reminded me of it in a late-night call once they got back to their hotel room.

"Mom, are you ok?" he asked. "I saw that man kick that door."

I giggled and said to him, "Yes, I am ok. That's normal around here. He was upset because a soldier got back late and missed accountability formation."

He said, "Yeah, I saw him try to sneak back in."

"Well, as you've seen, we don't get away with much. So it's easier to do what's right." He and his sister both serenaded me with "I love you" as I told them "I'll call you guys early in the morning. I love you both, too. Help your grandmother out and have a good night."

I spoke to my mom and made sure she was resting with ice and elevation on that injury, thanking her repeatedly for being here for me, and letting her know I loved her, too. I knew my times for movement were scheduled before theirs so I wouldn't get a chance to see them in the morning like some soldiers who would have time to wait around; but my flight was out early because I was heading to my next duty station for AIT (Advanced

Individual Training) at Fort Sam Houston in Texas. Here I come!

CHAPTER NINE
My Incarcerated Life

Vandrick and I lost contact for a bit which was to be expected when I moved from Columbia, South Carolina and got settled in San Antonio, Texas. I had a new reception center and new Cadre to acclimate to. They were no longer called Drill Sergeants with the large green hats. Learning the ropes here was a bit easier as the discipline was still ingrained within me from basic training. Some things, no matter where I go throughout the military, will never change. There was still the possibility of mass punishment and disciplinary actions, but not as common because we were supposed to be squared away at this point. My mom and my children made it back home safely. My mom definitely needed to see her Podiatrist because her ankle took a beating on the trip. My heart was still broken about the reality of my son's fragility. That will never escape me, seeing the tears streaming down his face. I definitely needed to have a conversation with his dad about it so we could try to figure something out to lessen the blow of the next year and a half ahead of us that we still needed to get through.

Luckily out here, the one restriction from basic training was over. I now have access to electronics. That was one thing that my mom did bring to me at graduation: my laptop. I needed that. No more locking our cell phones and computers away in a separate secured area. I can now keep it locked up during the day in my wall locker or even keep it on me, if no restrictions were imposed on a particular day. My calls would be less restricted than they were in basic training. Maybe having more access could help our son a bit more in dealing with his separation issues. With the guilt I felt, I was willing to try anything. I sent a letter to Vandrick for his immediate attention to the last known address I had for him at California State Prison, Folsom. I could only hope that nothing's happened and he is still there. We need to collaborate on our son's behalf. All I could do now was to wait. We were reduced to the delivery timeframes of the United States Postal Services.

Thanksgiving was soon approaching and I would be here on post. This was the first time I had ever been away for a big holiday that we celebrate as an entire family, both immediate and extended. This was definitely different for me. I will have to rely on the internet and phone calls to wish everyone a Happy Thanksgiving. I wondered what was done here on post to keep the morale of the soldiers up and festive without their immediate family to interact with. Before long I'd find out. They had a feast in the DFAC and the Sergeant Major came and carved turkey so that we all would recognize that this group of people was just as much our family as the ones we'd left at home. As I recall my reserve unit's holiday party the kids and I attended last year, there were no limits to the amount of food we could eat. They even brought out crab and lobster tails.

However, what you quickly figure out is when you're as active as we were with daily PT and height and weights measured regularly, there was no way you wanted to make multiple attempts at eating second and third helpings of any meal, even if we had all the trimmings. They really went all out. Some Cadre dressed in their Class A's and not in their combat uniform. It was nice to see. They made it a festive occasion as best they could. I still missed my family and as soon as we were able to get back to the barracks, I needed to make calls, send texts and make Facebook posts to let everyone know they were on my mind. It was nice to hear their voices. It seemed like it was a normal and traditional dinner for them as the others in the past except, I was not able to be there.

The following Sunday will be my daughter's 5th birthday. She and her dad decided to put a party together at Chuck E. Cheese. I wasn't thrilled about the location, but hey, I wasn't attending either. I did what I could from where I was at. My mom told me she wanted a bike so I made sure to send some extra money in addition to the monthly allotment that I already send for my children, but this was specifically for her birthday gift. I would receive pictures later in the day of her blowing out candles on her birthday cake, her friends and family playing the games while enjoying music at Chuck E. Cheese. No matter how I felt about this place, just knowing she was smiling and having a good time was what mostly mattered. She and I were able to talk on the phone and she told me all about her birthday and all the things she had done to prepare and make sure it was a success.

Yes, my daughter was very self-sufficient and will take care of things herself. She just needed access to a credit card and she would take care of the rest. She knew who she was inviting and found all of their information, printed out the invitations, and mailed or personally delivered them all by herself. She was not going to wait on anyone when she knew for sure she would get it done. I laughed with her and I am so proud of her, but I was also a bit gloomy as I realized how much I am missing her growth through moments like these. But I didn't express that longing with her. I would wait until she was no longer on the phone. I told her to tell her dad, "Thank you" for making sure she had a party. It was not as big a deal to him for her to have a party every year, but he made sure she had his credit card so it was possible. I also told her to tell her brother, "I love him" and I would talk to them both again and actually see them really soon. Christmas was just a month away.

The AIT classes would not start a new cycle that I could attend until we were back from the Exodus break which runs from Christmas through New Year's Day. So, we stayed around the barracks much of the time throughout the weeks, performing random details during the day or lots of barrack's maintenance and cleaning where everything faces some type of inspection. It gave us time to have access to our phones so that always made it better to be able to stay connected to back home. Also, it gave us available time to finalize travel plans to get home.

One night I received that unknown number across my cell phone screen and it was a collect call from Vandrick. I had been anticipating his call, or a letter, so I accepted the charges. He greeted me with a "Hey stranger" and likewise I did the same. He told me he recently received my letter and because of its urgency he wanted to try his luck at making this call so he was pleased that I was available to answer. We cut past the small talk and got right to the topic of our son. It was long overdue. I shared with him the details from the graduation and the tears that our son found hard to contain and my feelings. I told him he needed one of us. Vandrick agreed with me, and it is at this point where I am able to find out that he had made attempts to get our son picked up to visit him and was being blocked from the smooth process that had been established before leaving home.

He also told me he knew that the calls he was making to our son were no longer being answered and he was only getting voicemail no matter when he called. The last time he was able to really talk to our son, he told Vandrick that he wasn't as comfortable as he hoped he'd be; he felt he was being picked on. That's when Vandrick invited him to come and visit so they could talk about it more face-to-face. Shortly thereafter, he wasn't able to reach him anymore. Vandrick felt he was being pushed out of his son's life, something he knew would never happen if I were at home. I listened attentively and I am frustrated because, of course, none of this is being told to me by my family. There hasn't been another issue with my children brought to my attention to be addressed, aside from the one call during basic training where there was trouble with my daughter. I assured him I would speak with my family about this matter and we could discuss it further on another call since his time was running out. And I also had to head down to formation and get ready to march off to chow. I am spending time with not much to say to anyone, as my thoughts were now overcome with the question, *"What else do I not know is going on?"*

I knew I was isolated during basic training, but that is not the case anymore. I have been accessible via phone and laptop for a while. Why is no one catching me up or even letting me know if things are going wrong? I am definitely out of the loop. I needed to finish eating, march back to the barracks, perform whatever duties remained for us to do, so I could get back to being able to make calls. I have got to get some answers to my concerns for my children. Once I was settled in the barracks, I called my mom and she answered the phone. She was normally the source of information about what was going on. The kids have no issue with talking to her, so if anyone knew anything it would be her.

"Hey mom, how's it going?" I asked, cordially.

She replied, "Hello, dear. I'm doing okay, and you?"

I don't want to take up too much of her time. Plus, I never knew when I was going to be called away to do something else, so I kept it short, sweet and to the point. I told her how I felt about seeing my son in tears and I knew the time was wearing on him not being able to see me. Then I told her I just found out he was also being cut off from seeing his dad when I made every provision possible for this not to be the case. I am also getting information that they can't even talk on the phone to each other. I told her he mentioned to his dad some things going on with him and his dad made time and a way for him to visit and then that's when they were cut off from communicating. I finally said, "I'm trying to find out have these things really taken place or is this exaggerated information?" My mom didn't go into specifics for the most part, but she did let me know that my son had mentioned some things he didn't like; that he felt like he didn't fit in as we all had hoped. She said there were even a couple of fights that took place. As for making sure he could see his dad, she'd given me her own opinion regarding that.

"I don't think him seeing his father is a good idea. When he comes back from visiting, he isn't in the best of moods. Plus, there are other things he could do with his time."

Let's just say I was not feeling what I'd heard. This only made me feel worse because I could neither return back home nor get my kids and bring them out here with me during this time. There was a little more freedom at this phase, but not enough to bring along your family. This decision to separate my child from his father for whatever personal reasons was not ever a part of my parenting plan. As long as my children fathers' pose no threat or danger to the wellbeing of their kids, I was okay with them having whatever relationship they desired to have with them. And I know that even though prison wasn't the most ideal environment for any person to be in, I knew his father would never allow anything to happen to any of his kids, just as I wouldn't. And I am also certain that our son needed access to at least one of his parents and currently I couldn't be there for him. This conversation also further proved why I needed to hurry up and get my butt through these next three levels of courses successfully and get myself back home. As long as I was out here, I had no control over the things I intended for my children. I had to just go with the flow for now. I made sure to thank my mom for talking to me and appreciated the details I did receive, and ended the call by telling her I needed to get back to taking care of some things on post so I would talk with her later.

It would be a while before I could sit down and write Vandrick again. Plus, I needed some time to think and sort out my own thoughts and feelings. I needed my letter to not have all of my emotions poured out onto paper, but instead, contain well thought-out details of how all of us will be able to get through this time I have left to do. I would be able to compose a letter eventually, but unfortunately, I would have to tell him I was not in California so I couldn't control the things that were currently taking place, even when it involved my own children. This was so despairing for me to admit. I wished I could say something more assuring, but for right now this wasn't the case. I'd mail this less than desirable letter off and wait for a later response via phone or mail. Hopefully the response is by phone because the days were getting close for the post to reduce to skeleton crews and no mail would reach me until we return back from the holiday break in January.

Thankfully, we have made it to the Exodus break from Christmas through New Year's Day. I'm heading home to spend roughly two weeks in California. We double-locked our lockers to make sure they were secure before we departed. We were released close to midnight to go to the airport with our leave records in hand, itineraries planned and ready to go, yet all of our flights were not departing until the next day. This would be the first time I'd ever fallen asleep in an airport, but I was happy to do it. I was one step closer to being home with my children. This was my only focus; not how uncomfortable my accommodations would be for the next few hours. I was not making this visit for anyone else. I have missed my children terribly. Whoever I see outside of them was considered a bonus. This trip was strictly about me and my children's time together. As a matter of fact, since I already had a rental car lined up, I made arrangements to isolate us and go away for Christmas.

We traveled and spent our time in Las Vegas at Circus Circus. This gave us some time to reconnect, but also allowed them to change their routines and have some fun as well. In the past, my Christmases were spent shopping and getting them everything they could think of, as most parents do around this time of year. And the kids, a few hours after opening every gift under the tree and me acting as if I hadn't just gone into extreme debt, would say something ridiculous like, "I'm bored!" Now, having spent so much time away from them, I have learned something for sure—my time with them is more valuable and memories last far longer than any gadget a manufacturer could make. To just be near them would make my entire holiday enjoyable.

Vandrick and I had been able to talk a few more times. He was happy I was home to spend some much-needed time with the kids. And unfortunately, with only 2 weekends to do it in, I won't have time to take our son for a visit. But he had a few plans himself. So, he won't spend the holidays alone as well. I let him know about my plan to go to Las Vegas and his only request was for some pictures of me and the kids together. I was willing to accommodate that request. I had also built up the courage to talk to him about the discouraging news I received regarding my family's position about our son visiting. We were not pleased with it, but we also knew our hands were tied in the matter. My family had the upper-hand and control in this case, although temporary. We would have to take the blow on the chin and invoke our parental rights and restore order once I got back home for good.

I do appreciate the fact that Vandrick didn't get angry at the things I was saying to him. He did understand that I was between a rock and a hard place when it came to certain topics because I was not there. He didn't put me under more pressure than I could handle. He is disappointed, certainly, and I could hear it in his voice, but he was also compassionate towards the situation. I needed that from him, to ease some of my own guilt in having to be away. Being in prison with the restrictions placed on his life, he could truly empathize with me in my situation as I try to adjust to this version of my incarcerated life. It was a constant reality that he faced daily. The similarities in our circumstances made it crystal clear: prison and the military had identical model structures.

While talking to him, we still hadn't spent any substantial amount of time discussing the ins and outs of his case. I don't know if we will ever get around to that. It's been a few years now that he has been locked up. He spoke optimistically about different things he was trying to do to get out, mainly trying to set aside enough money to hire a good lawyer to fight his appeal. I don't ask many questions about how he was acquiring the money. I've never asked him questions like that and I was not about to start now. In my experience, the more you know the more you are responsible for knowing. I have determined that I don't want to be held accountable for anything outside of my control. I want Vandrick to be safe in prison and since he said he didn't commit this particular crime; I would love for him to have the opportunity for that to be looked into.

Any time I have ever spent inside the visiting room of a prison, and with the underlying tension and energy that is present, I know it was not a place I'd ever want to be forced to be there long-term, especially if I believed I was there for something I did not do. That would drive me extremely close to insanity. I don't think I would be able to spend one moment resting until I could figure out a way to be free. I could still hear from Vandrick's tone of voice that he is going through some things, deeper issues, but I don't know what to say to break the surface that would cause him to open up to me, but I have patience. I will give him the necessary time to build up the courage to say whatever it is that is troubling him. I don't want to push him too hard. His conversations were very light-hearted and he was always concerned about us, but I knew without his mother he must be going through a tremendous amount of internal suffering. Maybe that was the stress I could hear when we talked.

Vandrick is now a participating member of the Muslim community and he seemed to be allowing his religion to comfort him in a way no person can. I could pick up on this from him because he was not breaking under the pressure, even though he was masking his hurt to some degree. I was grateful that he had found a religious practice that he could depend on through this time in his life. Aside from any foolishness that might be going on, he is gradually maturing into a better and more accountable man. Maturity takes time and is a reflection of an individual's internal self with a commitment to changing flaws that were recognized. And even though he is not fully there, to know that he was working on the steps was encouraging to witness. His children have a great stake invested in him, making the needed changes and hopefully someday, benefiting from him being home. Therefore, he would be there to guide and help them not take the same path of being in the streets and ending up in prison, or worse being killed.

The kids and I have spent our much-needed fun time in Las Vegas and now the time was winding down quickly for me to get back on the airplane and head back to Fort Sam Houston, Texas. I find myself crawling back into my shell. Normally, for New Year's Eve, I would be at my daughter's father's night club, helping him out for the night. Thankfully, he no longer owns that place. Or, I would go over to my church's New Year's Eve celebration. But for the first time in many years, I don't want to go anywhere. I attended church services the first Sunday I was home. And, I am currently not in the mood for any type of celebration. So instead, I was at the house in the old neighborhood, spending my last moments with my kids.

I have washed clothes and hesitantly prepared my bags to head back on post. My mom and my cousin from down the street have gone to the church services. I reminded them to be safe because this area has not changed and the neighborhood would be the same as it had been every other year once the clock strikes midnight. There would be a barrage of bullets with no names attached, and illegal fireworks filling the air. But what the participants failed to remember was that old saying, "What goes up must come down!" We just prayed that no one was hurt from all the careless displays of celebration. At times, I find my eyes filled with tears and at other times I must remind myself to smile because my children were present.

We watched "Dick Clark's New Year's Eve" special as we lay down close to the floor. Sitting or standing up normally was never a safe idea during this time of the night. Even the church's Bishop knew not to let a single member out until after the shooting had stopped. The church was located just a few blocks over from our house. It was a well-known mega-church in the area and I have been a member since I was 14 years old. It was located a little further away when it was a much smaller congregation. Internally, my thoughts were forced to roam everywhere as I am trying to encourage myself that I could complete this next phase. I was counting the days and weeks backwards because somehow it made me feel like the time was not as long as it actually was. I have some battle-buddies back on post that I know will help to make the time feel less dreadful. I am encouraging myself with every thought I can think of to keep the depression from settling in and finding residence. But all I seemed to be able to think about was, "Oh no! I am leaving my babies again." I have decided if there was one word in the dictionary that I could remove, it would be the simplest of words and the hardest for me to accept—goodbye. Why must it exist?

I made it back on post and I was ready to get into training. There was a lot of medical training I was about to receive in a short amount of time; the pressure was on for sure. Either you got it or you didn't. There was an undertone to this course that suggested if you didn't pass you would be reassigned to the needs of the army. So, I've got to perform well and prepare to move on to the next phase. It was time to clear my head of all unnecessary thoughts and buckle down to what was required of me daily. Of course, there would be long days and nights, school, physical fitness, simulations and field exercises. Then there were repetitive theories to make sure that everything was ingrained within me and could be reproduced efficiently through muscle memory.

During downtime, I would take the opportunity to rest. There was a place not too far from the post that many soldiers patronized called The River Walk. I have gone there before and it's got lots of restaurants and shops, definitely a place to go at least once as a tourist, but hanging out there regularly didn't interest me much. I'd prefer to talk to family back home, catch a few movies and enjoy the rare occurrences of peace and quiet in the barracks. With four open bays stacked with females and 20 per living quarter, quiet was a rare commodity unless everyone was asleep and even then, it wasn't completely quiet. You had females in and out of the bathroom, the fire guards talking as they pulled their shifts, or the Cadre, the sergeants, roaming through and checking bunks with flashlights to make sure no one had escaped. Or, there were those like me from the west coast where the hours were not ideal to talk to people back home due to the time difference. So, either I would text to keep the sound to a minimum or find a place where I could speak at a whisper if it was necessary for me to call. Yeah, there were a few of us that didn't get much sleep trying to catch up on calls, including and not limited to late night booty calls. Hey, you gotta do what you got to do to keep yourself and your loved one entertained. It's only awkward in the morning when I'd get the random, "I wasn't sleeping last night and that must've been your man you talked to last night." I typically blush with a simple, "yes" and keep it moving. No need to go into explicit details with my battle-buddies about what needed to be taken care of throughout the night.

As far as friends I would talk to my former high school sweetheart, but he had a few situations he was involved in so I was learning not to take him too seriously. However, it was cool catching up with him because he at least could relate to what I was going through; with him being military we had plenty we could relate to and he was a great source for advice. He often teased me about the new Army being so laid back from what it was like when he went through in the mid 90's. I had nothing to compare it to so I had to take his word for it.

But my more serious relationship was with a friend from the old neighborhood. He had been interested in me for many years. However, because of so many things I knew about him and the things he was wrapped up in, it was difficult for us to really put the effort into making it work, officially. In the past, we'd get close then I'd push back. I loved him tremendously, but I was also afraid of the consequences of his lifestyle. If I thought Vandrick was ever in too deep, this guy would put him to shame in how deep he was involved. I tried to convince myself that the distance could protect me a bit from being caught up in his madness. Plus, no matter what, we have always remained consistent about one thing—even if a relationship between us never evolved, our friendship was the most important thing to the both of us. I've known him even before I ever met Vandrick. I grew up with this guy. I think I was about six years old when his family bought the house right next door. Close friends would ask, "Aside from a handful of guys, why have you always ended up in relationships with guys that have a similar background of being in the streets?"

I tended to answer with one of two responses, "I don't think I would do well with a guy that wasn't at least a little bit rough around the edges. I don't need a guy who is more afraid of a confrontation than me. That would only make me immediately lose interest. I don't need him to terrorize the streets, but I don't need him to lack confidence either." The other response I'd give if I didn't feel like answering in great detail was, "Look at where I come from and what I see regularly. There are not too many nice guys to choose from."

Yeah, those types of guys come with a set of rules of engagement of their own. I have never been a judgmental person and I knew how to think for myself. But as a precaution, it was extremely rare for people to even know who I was dating, unless I wanted it revealed. So, it surprised many when I announced suddenly that my neighbor and I had gotten married. It had been discussed even before I left for basic training. It happened so spontaneously because I had to work around the training schedule in order for it to even occur. I was able to pull it off with the help of a battle-buddy who also knew it was going to happen, eventually. I broke the Golden Rule told to me by many soldiers before I ever left for basic training: Don't get married in AIT! But I did try to somewhat follow the rule and didn't marry anyone within AIT. He was from back home and it was already in the plans. He was who I was spending my nights with, engaging in intimacy over the phone. So now they are classified as booty calls with my husband.

Vandrick and I talked again. Our letter writing was becoming less necessary and phone calls were easier to schedule. I let him know I was married, too. He congratulated me and seemed sort of surprised, but not really. He was used to my spontaneity and oftentimes discrete actions. He asked the standard questions, nothing too invasive since he already knew the person I was married to. He asked, "Are you happy?"

I simply replied, "Yes, I am."

"Does he treat you and the kids alright?" was his next question.

"Yes, he does. He asked them if he could marry me before I even left for basic training."

So out of the blue he asked, sarcastically, "So, you decided not to save yourself for me?"

"You are married, silly," I replied with a laugh.

"I am Muslim and I can marry more than one woman," he stated rather arrogantly.

"Now you should know that is something I would never agree to," making sure he understood *that.*

We laughed at each other. I never took any of his comments like that too seriously. I am so used to our flirtatious banter that it came off as normal and nothing to put any real thought into. We continued to discuss the latest regarding what was going on in our lives. He was curious to know how I was holding up in the service and I wanted to let him know I was still concerned about him in prison. We were both likely playing down our surroundings and situations by making it all seem easy. We didn't feel there was any need in raising alarms if none were needed. To be honest, I was really doing okay. I have found a pretty consistent routine and I don't hang out too far beyond the barracks and only eat out occasionally. My studies were going well and I was passing all my requirements. I still have my moments of battling with the same injury from basic training, but that was why I made sure I took my downtime and rested up. The training could be rough on the body, when you were not at 100 percent. And, this Texas heat was nothing to play with either. I just mentally had gotten to the point where I basically told myself it was mind-over-matter and positioned myself so I didn't cause even worse injuries and told myself that once I was home, I would have the time to deal with it all. I would not stay any longer than I had to, trying to doctor on anything. My children and I have set a date for all this to be over and I am going to honor that even if it kills me.

Vandrick was sounding about the same as I tried to gauge our conversations. He was still sticking to small talk. He was willing to share a few of the unbelievable things going on in the prison with me as he became acclimated to his new life of craziness, but he still hadn't mentioned much about his case. Vandrick has visited with his kids only a couple times, but not nearly as often as he used to with his first wife. I didn't know what was going on with that, but I knew it was troubling for him. His intention was always to keep his kids together as much as possible, especially during his prison sentence. If they couldn't come to visit him, he liked them having slumber parties and days together. He wanted his kids to grow up together and know each other. He enjoyed his time with his kids and any woman that didn't know that about him, and respect that part of him, wasn't paying close enough attention to who he was and how he was brought up. He might have been a guy that represented the streets, but when it came to family his roots were deep from his upbringing in the South. From the undertones of his conversations with me, I knew I had questions about what was really going on with him. But as usual, I was not going to dive too far into his business. One thing I knew was that he had requested to see his kids. I could say it with all certainty because even though my family placed their own ban on our son, it hadn't stopped him from asking me if I could try to get them to let him visit. So far, my efforts and his request had fallen on deaf ears.

I received a few unexpected gifts and calls from some friends; my co-worker who promised he would not forget about me lived up to his word. I appreciated the cards and even flowers I received from him, a group of other co-workers, as well as my husband has done the same. I can be in the barracks, minding my own business, and get a call to come down to the sergeant's office. Or, as we are marching back from class, I have been met with surprise bouquets of flowers and candy. I still enjoy the mail I am receiving from my children with the packets of graded homework and artwork. It was refreshing to know that I hadn't been forgotten about in my extended absence. I even received an unexpected letter or two, checking in on me from Vandrick's former, longtime cellmate at California State Prison Salinas Valley, the one that he studied Islam with. Vandrick left my contact information with him when he was being transferred out, so it was approved by him that we could communicate. He wouldn't have my information if it wasn't someone Vandrick trusted.

I think the prisoners understand the importance of being able to connect with life outside of their immediate surroundings more than some of my family and friends I have left at home. They are definitely more consistent in reaching out to me throughout the time I've been on this journey. I make time during some evenings, and more often on the weekends, to respond to all the mail and gifts. I always made an effort to reach out and let everyone know how much I appreciated their support. It kept me going on the rough days, the days where I'm questioning myself as to why I'm even here. Those days were days when that push from my support team was most needed. Things such as Facebook helped with reaching out as well. That source provided the immediate boost of energy when you started thinking it was time to pack your bags up and run towards home. A quick, "Hey, we're proud of you," joke posted or a "Girl, keep going," reminded me quickly that whatever challenge I was facing in that moment was not bigger than me. I could only hope my heartfelt gratitude was interpreted in the manner in which I sent it. I couldn't reach out and hug anyone, but I sure hoped they could feel as if I did.

Well, it was coming close to the end. I had made it through another phase of training. I couldn't believe it when I looked back on the many things we were expected to complete; the craziness and people I had to adjust to. But somehow, I am at the last field training exercise of the cycle and once we get back there would be nothing much left to do except clean and turn in our gear and prepare for another graduation ceremony. I find myself amazed at the progress and accomplishments I have made and the strengths I never knew existed within me that I have tapped into. I can honestly say that I was growing as a person throughout this experience. I have stayed mostly to myself, but I have become friends with a few very fun battle-buddies. I have gone shopping a few times at a faraway shopping center and out to dinner at a few nice restaurants, too. I don't like doing much too close to the post because more than likely I will run into soldiers or Cadre on my free time and I prefer not to do either. I had a taxi driver that I could call directly and he was used to my favorite request to take me where I wouldn't run into anyone from the post. He knew I would pay him for the extra mileage since we wouldn't be close enough for him to run back and forth to pick up more soldiers traveling closer to the post. So, I always tipped him handsomely for the money I caused him to miss out on. But it was worth it for some of the places I have been able to see.

I valued my away time tremendously. I have one battle-buddy that I am always with. She is the same one that was with me when I got married. She is similar to me in temperament so she doesn't mind when I tell her I don't want to go anywhere near the common hangout spots. I don't want to break the rule of not having a battle-buddy with me at all times, so in order to leave she had to agree. Thank goodness there wasn't much convincing needed for that. We had occasionally cut it close on making it back on time, but somehow, someway we never seemed to miss a formation or even a couple of unexpected recall formations. We never came back hammered or drunk. Neither of us were heavy drinkers. We just liked to see different parts of what San Antonio, Texas had to offer. I didn't leave post often during the school cycle, but the times that I did were considered a much-needed break.

This last week I decided to change up my routine. I wanted to let my hair down and have some unadulterated fun. It was noticeable enough that I had another battle-buddy ask, "Where are you going?"

"To go do some laundry," I jokingly responded.

She looked at me and asked, "Where is your laundry bag?"

I laughed and replied, "My battle-buddy must have it" as I scurried down the stairs, dressed like I was going anywhere except to a Laundromat, and ran across the street to my waiting taxi cab.

During our last field training exercise I met some friends that would extend beyond AIT because my next phase after graduation would still keep me at this same post. My battle-buddy was headed to her next duty station along with the other "battles" I had become close to. Everyone had their orders in hand where they were required to report to. It was only about two of us that would just move to the other side of the post for our next phase which was continued training. So, the opportunity to meet some new people that were stationed here helped me to transition smoothly. This graduation granted me a new set of rules as well. I was no longer a high risk to the Army. I was now officially considered "permanent party." I didn't need a battle-buddy to go wherever I went. I could move around on my own. I didn't know how much I missed my independence until it was given back to me. Not that I didn't enjoy the companionship, I truly did.

My battle-buddy and I were definitely kindred spirits. She was older like me, married and she had children back home, waiting on her return as well. That common bond connected us and gave us both a deeper understanding of each other. I knew I would miss her as we parted ways. No one in my family attended this graduation since I wasn't leaving. It wasn't necessary to pull my kids out of school and their daily routines for just a few hours of recognition. I paid for a video of the graduation that I could share with them later. However, if I made it through these last two transitions—Phase One and Phase Two of the nursing program—they would definitely be attending my final graduation. That is the graduation that would send me home with certifications in hand, reporting to my reserve unit as a Licensed Army Practical Nurse, mission complete.

CHAPTER TEN

2 Grads and a Divorcee

I have made it over to my new location, still on the same post, but on the opposite side of the grounds. I'm no longer in the AIT area. My cycle was not scheduled to start for another week so I was one of the first to get there and must wait for the other soldiers to arrive. I was sure I would be kept busy with details and things to occupy my time, but when I wasn't, I had some other business I could handle so when classes started I would not be juggling everything all at once. I would use the time I had now wisely. I would still have to wake early and do PT, physical training with the rest of the unit, shower and be in uniform just as they do. They would leave for class and I would remain back at the barracks.

There were a few people here that were being sent back to their units for not passing their test or for other reasons so I would be with them as I waited for my upcoming start date. The barracks were two-person rooms and with me getting here so early ahead of the rest of the class, no one had occupied the second bedroom where I was, so far. After being in the larger bays of the barrack's living quarters with multiple females in close quarters, I gladly appreciated not having a roommate for the first time during this entire experience. However, that won't be determined until everyone is here. I wanted to enjoy the solitude for as long as it lasted. The other thing that was different from what I had been accustomed to at the other locations is that you are authorized to drive your own vehicle. I would have loved to have my car with me, but it wasn't in the plan from the beginning so I resigned myself to go on without it. I would overcome this setback by renting cars or using my handy taxi cab when I needed to get around and off the post.

During the day, I had free time and the only restriction was that I would be able to be located quickly if there was a detail that needed to be completed, and that I would not hang out all day in my room in the barracks. There were not many other places we could be so it was easy to be readily available. This also gave me time to stay in contact with my family back home. Contacting my kids and my husband were always at the top of my list, but I also found time to speak to other family and friends to just catch up on the latest. Vandrick and I were also continuing to keep our commitments to one another to stay in contact by whatever means were available. We would have moments where we'd lose contact for a brief period of time, but rarely that we'd go too long without catching up. We did our best under the restrictions we both were under. His situation was far stricter at this point than mine, but very similar when we viewed it from the lens of free-will versus freedom.

I believe some of the communication lapses that Vandrick and I were going through now had little to do with lack of time because I am more available than I ever was at my last two locations. I felt it had to do with whatever he might be occupying his time with. When we did talk, he mentioned more often than not how the prison term is weighing on him and how he was doing what he could to try to come home. The details were vague, but the information he did share was filled with tension. He was up to something and I knew his goal was his freedom. One thing I knew to be true about Vandrick was that when he put his mind to something his intentions were to see whatever it was all the way through. I just prayed that he was safe and didn't get so caught up in his plans that he failed to clearly see every step he needed to take. Falling over your own mistakes was a tough challenge for anyone to face.

I don't hear him speaking much at all about programs the prison offers or if they had anything in place to assist inmates in combating the system. When I looked at it from a business perspective, wondering what financial gains would be produced by the continued confinement of inmates, it made sense why resources and accessibility would not be promoted heavily by these prison facilities. It would have to be a personal fight and a self-motivated battle in order to wake up daily to deal with the prison politics that would never fade; then have enough energy left to fight the additional battle of freedom by challenging the judicial system for a wrongful conviction. That is a hell of a war to fight, especially when you weren't equipped with the tools necessary to even understand all of the regulations and laws previously established without taking into consideration an inmate's needs and lack of understanding. I didn't know if his current wife and the people he had around him were efficient in this area, but I hoped that as he communicated with them they were able to get into the trenches with him and help him sort through all the things that were required when dealing with legal defenses and appeals. It definitely was a battle that took more than being street savvy and some money.

Vandrick had also made it obvious to me that he was taking his religion more seriously. I was grateful to hear that. I looked at it as he had support he could utilize as a source of checks and balances. We all needed something greater than ourselves which could remind us that when we are falling short, how to recover and move forward. I was not completely sure how any religion was practiced inside a prison. I was led to believe from the strict policies I witnessed as a visitor, it was likely very challenging to practice anything in a large group setting out of fear that something might go wrong, and quickly. I pictured in my mind that the inmates were attempting to pray and there were at least two CO's standing and monitoring the perimeters, ready to subdue any escalation of violence immediately. It all seemed a bit intimidating and less relaxed than a peaceful church setting, but it was what I envisioned when I thought of any congregation of people inside a prison facility. I had never studied Islam, but I knew it had a structure that was a strong foundation and I also knew that it was not too far to the left of my religious practice of Christianity.

Being in the military and with the current state of the world, I was becoming more aware of Islam as a religion and that Muslims were being discussed more frequently than ever before. We were in the post-911 era. In the media, it was definitely portrayed more negatively than positively, but I had never gained knowledge solely from one source of information. If that was the case, I wouldn't even understand my own culture. So I took bits and pieces of what was stated in the media and researched the rest on my own. Eventually, with all of the information I had discovered and combined, I could come to a rational conclusion, eliminating the stereotypes and subliminal messaging of discrimination.

I am not sure what is taking place, but without warning my next contact with Vandrick had changed. He was moving to another prison. I was moving through my classes and phases as well. I was now acclimated to the military lifestyle. The classes had been taught in a rapid and strategic method and if you could keep up and retain the information you would do well. If you got sidetracked ever so slightly you could easily fail. So, I tried to remain focused and honor the commitment I made to myself and my children. When it was time to learn I wouldn't let myself be distracted and when it was time to have fun I would do that, but I would never combine the two. The military had forced me to build a routine that I strictly followed. It was a structure like no other schooling I had received.

The other soldiers I was waiting on for the class cycle to begin had made it safely and some had built study groups. I had attended a few times, but I did better wrestling through the material by myself. It had helped that I didn't gain a roommate in the process, so my two-person barracks was my solo domain. Although I was all alone, I didn't sneak anyone in to entertain. On days that I wanted to enjoy some company I would go off-post, now that I had gained that daily luxury. The friends I met during the FTX (Field Training Exercise) I did get some time to see on days where my studies were not too overwhelming. That interaction was necessary for the time being. It was nice to feel as normal as possible. I was able to take a trip for the Fourth of July holiday up to the Dallas, Texas area. In order to take a trip like that, paperwork had to be in order and for that much needed break it was well worth it. I rented a car and went to see my high school sweetheart's parents. They lived in Texas and I promised them that one day while near them I would make it a point to visit. I hadn't seen them since they lived in California when I was in the relationship with their son.

The breaks away from the daily routines were always refreshing for me, no matter how short lived they were. The class cycle ended and the timing couldn't have been better planned for me. I had a break between Phase-One and Phase-Two of the course since the cycle would consist of our class moving post, and some soldiers would be moving their entire families for this last phase of our training. So there was a need to provide a break so everyone could get from Texas to our next location. This had given me time to surprise my kids with a trip to California. It was in July and a day before my son's birthday when I landed at the airport and I couldn't pull it off successfully without help, so my mom and sister knew I was soon to arrive, but I swore them to secrecy.

I made it to Oakland on the 13th of July, came right into town and started running around in preparation for surprising my kids. My son was working a summer job as a camp counselor and my daughter was attending it so it was easy to surprise both of them because they weren't too far from each other. I went that day and bought my son his first cell phone for his birthday, a cake and balloons. I stayed overnight at home in the old neighborhood with my mom. She took me to drop off the cake at my sister's house and then we headed to the park where both of them were as their day was ending. With balloons in hand I ducked down inside the car.

As planned, we pulled up to surprise my daughter first. She was shocked and very excited so I would say we surprised her. She was normally the one to figure everything out, staying close to the adults, hearing the most, but this time I was able to pull it off without her knowing a thing. She saw the balloons and knew our next stop was on the other side of the park to surprise her big brother who had no time to get a hint from anyone. My mom and daughter got out of the car and got his attention. He could see the balloons filling the car, but he was unable to see me. I got out of the car and stood just a few feet from him. He turned and was completely taken by surprise that I was standing there with him. He rushed towards me as I opened my arms and he gave me a big hug with a huge smile on his face. Thankfully, this time he didn't have a face full of tears.

He was shocked that on his actual birthday, the 14th of July, I came home to wish him a happy birthday face-to-face. I couldn't stay too long in the Bay Area, but I did spend a few days with them before I was back on a plane and this time headed to Fort Gordon in Augusta, Georgia.

This was my last leg of this journey and hopefully I would finish strong. The emotional rollercoaster of entering and exiting my children's lives had taken a great toll on all of us, but we did our best to make the most of the less emotional experiences. I was glad that I could make it into town for the short period I did. Now it was back to the airport I go!

Flying across the United States from the West Coast to the South, I got to enjoy getting on an extremely small plane that flew us through turbulence and to this airport that was about the size of a city library where I was picked up by one of the sergeants I am in the class with. She brought along her children and walked me through where we needed to "in-process" onto the post inside the hospital. She was pretty well-established and will be living in On-Post housing with her children as soon as they could assign her a house. She and I got to know each other at the previous post because she was in the barrack room right next door to me. With her years of military experience she was willing and able to teach me lots of things new soldiers would eventually learn over time. She was someone I could appreciate learning from.

I received my room number and went over to the barracks to get settled in. I arrived a day earlier than when we were scheduled to arrive so it gave me time to organize my room and rest from the long and less than relaxing flight. Initially, I was assigned a room with a roommate; but after a few adjustments I was moved upstairs. I believe they were using some information from the room assignments from the previous post, but they didn't need to accommodate all the soldiers because some were moving into on-post and off-post housing based on their family needs. Being a prior basic training soldier going through all the training consecutively and not having my family in tow, made it easier for me to just remain in the barracks through each phase of this process.

So, I was again without a roommate in my room upstairs just as I was able to enjoy a room by myself at Fort Sam Houston. However, I was prepared for a roommate if there was a sudden change in the plan. That's why I touched nothing on the other side of the room, except to dust and sweep to avoid any issues during an inspection. The barracks we were in had two sleeping quarters, a small kitchenette and bathroom. This was nothing like the days of basic training and AIT where I was sleeping amongst many and not a single private bathroom to look forward to unless I counted in the field when I had access to a portable potty. To be in a room alone was a luxury that I had no complaints about. Even if it was an error made in processing I was not going to make a fuss and bring it to anyone's attention.

I became acclimated to this post as quickly as all the others, but this one I will say was the largest of all the posts I had been on at this point. It felt like a small city within a city once I had the opportunity to take a tour around it. Here, I would split my time outside of normal military activities between morning PT, Change of Command ceremonies, morning and evening Flag Honors with nursing, clinical and didactic learning experiences. This was now my new life for just shy of one year. I didn't like being on a post of this size without a car, but it would be a bigger hassle trying to get my car shipped out here and then I'd have to drive it or ship it back to California when the cycle was over. So, I relied on transportation like I did on the post in Texas; taxi cabs and some really supportive soldiers that have their cars help out the soldiers that didn't.

When I want to spend a long period of time off the post I would go back over to the library-sized airport and rent a car for a week or two. Then I would return the favor and show my gratitude to the soldiers who were transporting me around by helping out others by providing some of the transportation. The classrooms and the PT fields were a great distance away from the barracks. The barracks were located closer to the hospital where I spent a large portion of my clinical time. The hospital was also where I picked up my mail; ate breakfast in the cafeteria if time permitted; sometimes ate lunch; and ate dinner only if I didn't have anything in my refrigerator that I could prepare. Some of our clinical experiences were also off post. Fortunately, we didn't have to think about transportation when we were assigned to those nearby hospitals. We were provided large, white military buses that picked us up behind the barracks and transported us to each assigned hospital, and picked us up in the evenings and returned us safely back to the barracks on post. And we had no worries about being lost or mistaken for anyone outside of the military because we were distinctly dressed in all-white nursing scrubs and freshly shined, black class-A shoes, and a few of us decided to purchase the less noticeable and more comfortable white nursing shoes. It was a well-established program and ran efficiently so it made the adjustment process easy to transition into.

After getting comfortable on post I could send my new contact information to my family and friends back home. Having my cell phone helped everyone stay in contact during the transition period that had become all too familiar. Mail was always a slower adjustment, but I made sure in-between getting used to my new routines I would sit and write to everyone that I had stayed in contact with so they would know where to send me mail. My family and friends were troopers as they were getting used to the transitions as quickly as I was. I always appreciated that first bit of mail that I received from anyone. So, with a smile I'd give a sigh of relief that the information I was sending out was being received and that they cared enough to take time out of their day to respond. With all the technological advancements that were taking place it was almost taboo to expect people to really sit down and do something considered old school, like writing a letter or mailing a card. Not when they could easily email, upload pictures, send electronic greeting cards, or text a message so much more quickly. I am so grateful when I think about the people in my life who remained consistent throughout this process.

During much of my time in Georgia I lost contact with Vandrick. I couldn't say what he's been going through. I knew where he was but our communication was significantly limited. Therefore, I was left to assume he had probably faced a few hiccups along the way because I knew he would have kept in better contact with me if it was at all possible for him to do so. I do spend time on the phone talking with his Muslim brother and he lets me know that there had been no word of anything major that I needed to be concerned about like a riot or something like that. That information calmed my nerves down to a minimum. I figured whatever was happening would eventually surface and then he could let me know what transpired. His Muslim brother always acknowledged how proud of me he was and my accomplishments as I've gone through this process. He told me it was a privilege to know a woman who was in such a rigorous line of work while knowing the potential harm that is associated with the profession. Based on the type of person I was before entering the military it was hard for anyone who knew me to believe this profession would be a perfect fit. But I pushed through it.

We often laughed about the number of hours I had learned to stay awake, especially since he considered sleeping his favorite pastime; he would tell me he needed his beauty sleep. There was a big time difference between California and Georgia so we based our communication schedule on when he would find time to check in with me.

He would tease me about my sleep deprivation during Flag Honors. This consisted of me finishing a round of studying throughout the night, falling asleep for about an hour or two then putting on my uniform and jacket because it was cold. I'd meet up with the other soldiers in the lobby of the hospital , go outside and raise the flag. Next, I'd head back to the barracks to rest for about 30 minutes before I either changed into my white nursing uniform to go to clinical sites, or stayed in the same uniform and went to class. This schedule was rough on anyone's sleep schedule but it was necessary for me to be able to accomplish all I needed to remain in this course.

This had been a hell of a process. I was learning a lot indeed, but I've had my days of struggling emotionally to get through because I truly miss things back home in California. Augusta, Georgia was definitely a different experience than any of the other places I have been. Therefore, to ease my moments of homesickness which set in periodically, I would grab a rental car and drive to Atlanta, Georgia. Up there things were a bit more similar to the things I was used to in California and luckily I had an uncle that lived there. Spending time with him helped to relax the anxiety that would build up from not being around family. The phone calls were helpful, but there was nothing like face-to-face contact.

My high school sweetheart was also stationed at the other Georgia post, Fort Benning, and so from time to time we would make plans to meet and hang out in Atlanta, too. A weekend getaway was always refreshing. It was the closest thing I had to normal. I was glad we had the opportunity to see each other when we could because not too long ago he was placed on orders to prepare to be stationed in Alaska. The unit there was already scheduled to be deployed to Afghanistan not too long after his arrival. This was his fourth deployment into a war-torn country. With him being an infantryman his number of deployments were rather normal. He had already faced three tours to Iraq and this was his first to Afghanistan. Nothing would make this process any easier, but we would see each other one last time when he comes to Georgia for a quick visit to take care of some last-minute details once he's moved to Alaska. Then he was going to be deployed for one year. It was rough for us both during this last visit together with the knowledge in the back of our minds about what could happen after he was deployed. He'd already had a few close calls the times before in Iraq. I witnessed the physical and mental scars he suffered from repeated deployments, but we made the best of our time together.

During our time together we laughed, talked, watched TV, went out to eat, and took quick naps. Whenever he noticed I was about to cry because my thoughts were taking over, he would say to me "Man, look up at the ceiling!" For him, he felt this would keep my tears from streaming down my cheeks. Or, his other playful statement of, "Swallow sand," in some way implied that my tear ducts would dry up. He knew his comments would catch me off guard. I would giggle and the tears would not fall as easily. He didn't want to make our time together any more emotional than it already was.

We had some big topics to cover in our short time together. We made our agreement that was always honored throughout his previous deployments: we will remain in contact. He gave me instructions of what to do if the worst-case scenario should happen. This was extremely difficult to listen to as a person who cared deeply for him. We have a history that was founded on love and for us to reunite after 10 years of being apart the topic of death was extremely overwhelming. I have been a reliable support for him in the past and he trusts that whatever instructions I am given I will see them all the way through without hesitation. God knows my prayers are that I never have to use any of this information that I am entrusted with. However, no matter what, I'll toughen up because I understand this was what real friends did for one another.

His previous deployment to Iraq was extended from twelve to fifteen months. Soldiers were facing stop-loss laws due to the historical intensity of two wars at one time. He became so aggravated because he felt he would never be able to leave, or in this case leave Iraq alive. We worked through those long days and nights, thanks to smartphones and computers. Thinking back on the past, I offered him encouragement such as, "we made it through before and we will do it again." I was accessible no matter the hour; I would follow the news, but not too closely because it got tough to watch sometimes; I would send care-packages every other month with items he needed to make him as comfortable as possible. I tell him to thank the transport truck workers for me as my packages filled to the brim always gave them a workout in that desert heat. Now that I am a soldier myself, I had an even better understanding this time around of the importance of having the support that was necessary to keep him pushing forward day-to-day. Our visit ended way too quickly with a kiss and a sincere farewell. The drive back to post this time was unbearable and my emotions were in shambles.

Atlanta had become my place to unwind, my freedom from all the constant PowerPoint lectures, clinical experiences, studying and weekly required testing. Our many classroom clinical instructors and Cadre were all awesome. Once we passed the formalities of getting to know each other and their need to lay down the ground rules to keep us out of trouble, they had been nothing but supportive of us. My class instructor and Cadre, a staff sergeant in the military, was not only a great source of knowledge about the curriculum, but also made himself available if we needed him for pretty much anything. He even let us meet his very bubbly daughter who would come to visit him occasionally at work because they didn't live too far from the post. Her presence helped to fill the classroom with plenty of smiles and laughter. I definitely needed the time the sergeant was willing to share because I was having a meltdown from missing my children. My motivation was dwindling because of the amount of time I was required to be away from home and my children. I was grateful for his open-door policy and pep-talks; it was very much needed.

Sometimes I couldn't hold back the flow tears when I was in his office and his previous experience during his deployment unexpectedly became useful in helping me as he told me how he had to push himself through that time away from his family friends and normality by simply reminding himself, "Sun up and sun down is another day closer to home." He also had no problem with advising me to step away from the books and be confident in myself with knowing the material that was being taught. He saw that I second-guessed myself often and could talk myself out of the information that I knew to be true. I have had to learn a thing or two about over-studying and test anxiety during this course. The amount of information that we must absorb and the timeframe in which we must reproduce it for the exam is intense. The instructors are compassionate. They proved they wanted to see us become successful in this course and our professions. I can admit I have experienced not passing two exams. However, what could get you dropped from the course and sent back to your unit without a license was if you failed the same exam twice. It was called a "double tap." I have not experienced that so I was still here and remained committed to the promise I made to my children. The soldiers that were in this course with me were all supportive of each other. In some aspects many of us were going through the same things, maybe not exactly, but we could at least empathize with each other.

When I have not gone up to Atlanta to get a breather, I have found myself going to church with some of my classmates, going downstairs to enjoy a barbeque in the covered bench area that had a huge grill, or enjoying a night out at a nightclub, billiards, or restaurant to eat something with the other soldiers; versus cafeteria food or microwave dinners in the barracks alone. When we first came on post we were the junior class and we had the chance to see the successes of our senior class as they received their recognition and then cleared the barracks and moved on to their various posts to start working as Licensed Nurses. Some will even remain stationed here where they will work at the on-post hospital named Eisenhower. And what I gathered from watching their experience was they made it through by building a tight-knit bond amongst each other, supporting each other as they could.

The torch had been passed down and now we were the senior class and welcomed our junior class who were now filling the barracks and moving into on-post and off-post housing. We did our best to make them comfortable and helped them acclimate to things here at Fort Gordon. I still did not get a roommate throughout this transition. I was happy, not that I wouldn't adjust to any soldier they assigned to my room, but it didn't happen and we had a lot of rooms because everyone did not stay in the barracks. I got to know many of them that were in the barracks or riding on the buses to off-site hospitals for clinical experiences. We all are receptive to each other as it is easy to adapt when all of our goals are the same—complete the requirements, take the NCLEX and pass. So, when Kevin Hart was in Augusta, Georgia for his nation-wide tour, some of us from the junior class and senior class all decided to go together and spend a night laughing at his hilarious jokes and headed from there to the after-party to enjoy dancing for the remainder of the night.

Finally, we have made it to the end of our cycle and it couldn't be a better time to be graduating in just a little over a week because a group of us planned a post-cycle vacation at the Black Bike Week at Myrtle Beach during the Memorial Day Weekend. We had plenty to celebrate because we all had passed that gruesome course. We all had different dates scheduled to take the NCLEX, but we at least knew we were successful in getting through all the tests the military required us to take. We could at least celebrate knowing that part. My squad leader had found a great place for us to stay and everyone going pitched in on the cost. We were on the 19th floor looking down on the beach and as far as the eye could see, the Atlantic Ocean. It was an absolutely beautiful view to wake up to for the next three days. It was definitely what I needed to see as I realized my journey in the military was finally coming to an end. I had made it from basic training all the way through to the graduation requirements of Phase 2 of LVN/LPN School. I never did any of it alone as I had family, friends, soldiers, instructors, Cadre and even complete strangers that I pulled support from. I am completely grateful.

The last morning I am here on Myrtle Beach, I wake and take a walk along the water before everyone is awake and just spent some very necessary time with God; this time was not about asking for a single thing, but to thank Him for being with me and providing me a strength I never knew existed inside of me. I had become proficient in so many different elements that I couldn't name them all, but I was proud of each experience and the fear I was forced to overcome. I felt more equipped and confident in myself than ever before.

As we all returned back on post, we continued with our military rituals of out-processing and making sure our flights and arrangements were in order for those of us scheduled to leave. I had already sent the flight itineraries out to my mother. She and my children would be present as I graduate this one final time through this process. This graduation will be different from the last one they attended because we would be flying home together! I would miss the soldiers I had come to know more as family than I ever looked at them as mere friends, but we knew that we must part ways. Everyone had gotten contact information and Facebook friend requests established so that as we moved on to do whatever was in our plans, we would remain as close as all this technology would allow us to be. We completed our much-needed packing and handed over the barracks to the junior class who were now officially the senior class and soon would be welcoming in their junior class. Some of our class had already taken the NCLEX as I had and I already had my results after an extremely nerve-wracking day of testing. I passed! I am officially a Licensed Nurse. I can say to my children, I fulfilled the mission that caused our long separation. Our time apart is not in vain.

My family were heading in from the airport and they would make it just in time to come straight to the auditorium where the senior and junior classes, family and friends were all in attendance for the graduation ceremony. It was not too long or too short, especially since I was on pins and needles thinking they might not make it on time. My back is to them as my family slipped into their seats just as the Colonel was announcing the honors and awards for our cycle. I was completely taken by surprise and felt I must have been in another world when she announced the recipient of the Nursing Excellence Award—it was me! Trust me, this was news to me. They knew I had no idea because they slightly laughed when it took some time for me to even recognize my own name and finally proceeded up to the stage and received the plaque. I double-checked and it was engraved with my name, PFC Qiana Riley.

I realized my family had made it through the cheers when I heard, "Way to go" shouted from the back and it sounded like my son and my mom shouting in unison. I can't see clearly as the lights were shining in my eyes and blurring my vision, but I knew the sound of family that resonated in my ears. I returned to my seat and the ceremony soon came to an end. After greeting my family with hugs, tears and laughter, we had refreshments. There were lots of families waiting to get pictures taken. Our class was the first class to make a request that was fulfilled to have one of our instructors from Phase One to fly in as our Guest of Honor, Mr. Neilson. This man was an awesome instructor who had a natural gift in his field for teaching. He took time to make sure we all would pass the course and he had a sense of humor that made the material stick and was far less overwhelming to absorb. There was no way our cycle would go without inviting him to see us all graduate. Thanks to him and a host of others, we can proudly say, "We did it!"

My family and I headed back over to the barracks. I had a few last-minute things I needed to do like ship home boxes of stuff I had accumulated. For basic training I flew in with one duffle bag; now I have a TV, radio, books and movies; and tons of new outfits, shoes and purses. Who in the heck knew I would collect so much stuff over the last two years? That didn't include all of the military stuff I had now, too. I needed to return my rental car because we were not flying out of that library-sized airport. We would fly out of the Atlanta Airport in two days. So I had to make sure I had everything as I packed up my mom's rental car with my overstuffed luggage and theirs. We would stay in the barracks throughout the night and then head to Atlanta to see my uncle before we left. There we would prepare for our departure back to California.

We couldn't miss a beat in our plans because my son was also going to be celebrating soon when we got back to California. He was graduating from middle school. He equally had as much to be proud of himself because he suffered through a rough patch just a few months prior and was pushed to get back on track in his grades. I am grateful to his core teacher who was instrumental in his success. Once she noticed his lack of interest in his work and saw that it was more emotion-driven versus a lack of comprehension, she requested a meeting. Thankfully, my Cadre-Instructor understood the necessary importance of my participation in being there for my son. I was able to attend the meeting via Skype to encourage him that this was not the time to give up on doing his absolute best.

It was not a day to be taken lightly as the conference was scheduled on a day when the Sergeant Major and First Sergeant were doing barracks inspections. So there had to be special privileges given to me for them to skip my room so I could participate in the remote parent-teacher conference. That one conference and my son completing a few make-up assignments and tests helped him to turn failing grades into B's. I am beyond grateful to all the people that made sure this moment was possible. At my son's graduation the emotions of what our family had endured with the amount of time I had been away along with the various things we had to overcome; it was only right that I was the loudest parent on the quad as his name was announced. I shouted out in complete celebration of my son's and our family's accomplishments and all that we had been through!

I was waiting to get news on where Vandrick was and what was going on with him. I am still in the dark about his status. I am certain he was not at the facility where he last contacted me. Now that I was at home I had plenty of things on my agenda to do to get myself and the kids readjusted. I couldn't spend too much time trying to figure out what was going on with him. He definitely knew I was no longer on active duty and how to find me. Housing, transportation and employment were the three major things on the top of my list to make sure were re-established. And as soon as I got an opportunity to do so I would need to get on top of this injury. I have endured the pain for a long time, but now that I was at home and under less pressure I could focus more on the healing process. Gratefully, I had a few months to make sure everything was all done. The kids were not in school, due to the summer break, so that made it a little bit easier on us all. I had to make contact with my reserve unit and in-process with them. This was important because they were also responsible for making sure I received my enlistment bonus which would definitely help smooth out some of the kinks in this transition phase.

I also needed to make sure I was up to date on my deployment status. While I was away there was information that suggested we were geared up and ready to be deployed. However, as long as I had been away, I preferred not to come home only to be sent right back out. Although I signed up for all of this, information such as this would take a great toll on my kids. When I inquired about my status, I was quickly updated and found out that most of the unit was already assigned and had already been deployed. The only need was for those volunteering at this point through other units. So, for the time being I was able to focus on the tasks I had already planned, except things hadn't bounced right back into place for me and the kids. It seemed more like everything that should have been easy came with its own mini-battle. It was frustrating to say the least, but I had never been one to give in during a challenge. So, I continued to chip away at everything that needed to be done until everything was all in order. I guess anyone who had been away for two years would face a definite adjustment period as well. The kids had their own agendas and itineraries that I was working into my agenda and itinerary.

As for my husband, I made the decision that I felt was best for the both of us which I was certain he wouldn't agree with, but a divorce was more likely the path we were headed down. Not that I wasn't willing to fight for our marriage, but there were some things that I knew I couldn't compromise on that had been taking place in my absence. It was easier for us to remain friends than for us to build up resentment towards one another that would be much harder to recover from. So, along with the other vital tasks on my plate I was making the necessary time to start the process of our divorce. There were no outsiders influencing my decision and I was not seeing anyone at the time and neither was he. This was a necessary adult decision that needed to be made to make sure that I and my children were my absolute first priority. My husband and I did not fight or bicker and I was pleasantly surprised that we were able to respect our initial agreement that no matter who wanted or needed to be released from the marriage we would not act like fools and cause each other hardship. Our friendship and mutual respect for one another had always been what was most important to the both of us.

I had finally heard from Vandrick. He was able to catch up with me despite all of my transitioning and moves. He'd run into a few hiccups during his silence, bouncing around from prison to prison. I knew our loss of communication was a result of something, I just didn't know what. He lost his cool and got himself into some trouble. A fight with a cellmate that he said was unavoidable. His points have gone back up so he was back at the higher-level facilities and even spent some time in, "the hole," a segregation unit. He seemed a bit remorseful for his actions, more so because he hadn't found a way to avoid his actions thus, allowing the circumstances surrounding him to be used to bring him back to those more restricted facilities. Not to mention how far away he was sent. He knew it would now be a greater hassle on his family to visit him. So, he relied mainly on mail and phone calls, again. I was relieved to know he was okay, considering the particular situation he had placed himself in, but upset as well that he couldn't remain doing as well as he was doing before.

The last I heard he was committing himself to getting his GED and participating in a few groups that were offered to the inmates. But who was I to judge? I couldn't say what I would do willingly or unwillingly if I were incarcerated. I just got a small dose of dealing with mass populations of people living closely together, some good people and some a pain in the ass. On a daily basis for almost two years I had to restrain myself from allowing the close quarters with so many different personalities to derail my goal of completing each phase successfully. This was not the easiest of tasks and I knew at least one particular time I was pushed beyond my limits. I had to calm myself from putting my hands on a sweet girl when she was sober, but a fool when she was intoxicated. She got so sloppy drunk one weekend that she peed on the floor behind my bunk in the middle of the night because she thought she had made it to the bathroom which was in the opposite direction of where she decided to squat and urinate. Fortunately, I regained my composure and had not allowed that incident get the best of me. So, if it could happen to me where soldiers are supposed to be held to a higher standard, I could only imagine what he had to deal with behind bars with society's defined criminals.

We changed the subject and I had a little bit of time to let him know that we were maintaining and trying to get ourselves settled back into life in California, but it was taking longer than I expected. He encouraged me to have patience and to trust that everything would eventually work itself out. I also let him know about my pending divorce. He was disappointed that we didn't last longer. Vandrick had always had a mutual respect for my soon to be ex-husband. Then he told me that after one of our detailed conversations about my marriage when I was stationed in Texas, he had a feeling that if things didn't resolve from that situation, he had a strong feeling I wouldn't remain married very long. He applauded himself, thinking he knew me so well. I guess he has known my intolerance for some things better than I gave him credit for, especially since the subject was one I had brought up to him to get clarity about. I didn't feel comfortable discussing the matter with anyone else. Based on the subject, I knew he was my best source for that type of information and I trusted he would have my best interest in mind and not beat around the bush with giving me his opinion. He gave it to me plain and simple and I respected him for his candid honesty. Well, our time was running low on the call and I let him know we would remain in contact now that I had information from him on his whereabouts. I also told him if my son and I could make a trip to visit we would, but under the current circumstances he had to understand that based on the distance it would be a definite strain to pull it off. He understood and the call ended.

My thoughts after that call were reflecting on the many changes that had transpired in just a little over three years. Some were great and some were less than desirable, but such is life! I officially had two professional skills under my belt with combat experiences as well. I gained a partial degree that I needed to complete. My children had grown two years in my absence and I had been married and soon would be divorced. I was in the middle of one heck of a transition that was teaching me the greatest amount of patience ever. I did acquire some new life-long friendships as well as lived in and visited parts of the United States that I never imagined I would see. This was only the beginning of this journey with my new MOS (Military Occupational Specialty). Now that I was a properly trained soldier there was so much more in store for me to do and see. I didn't train to get all this knowledge and skill not to utilize every dollar of it to the best of my ability for the United States Government.

CHAPTER ELEVEN
Life's Land Navigation

Much in life had changed, some for the good and some for the bad. I had found out that I was tapping into some things in my life that I never imagined I would face or even do, but I also appreciated and valued the woman that I was becoming every day. I had found an inner confidence that I did not explore before; quite honestly, I shied away from it. Now I was learning to accept the presence of my own thoughts and ideas. I've overcome lots of adjustments and challenges. The life I returned to was vastly different than when I left. The economy had taken a big hit and now financial risk-taking was at an all-time high as people were developing new ideas and resources to become successful. My children and I moved in with my father. I never saw this happening. However, I was grateful that he had overcome his obstacles with addiction and was on a healthy path in life, allowing him to be able to provide us with this option. I had the work requirements of two careers; I was enrolled in college again to complete my degree, and the cost of living in California continued to increase like an uncontained wildfire. This was the most likely alternative that made the most sense for us.

I am officially divorced and solely reliant on myself for all things the three of us would need. Also, after an emotional battle in family court that my daughter's father waged against me because he did not want me to move on with my life without him, I do not receive any form of child support for my daughter and I have never opened a support case against either of my children's fathers. So, with what I receive as far as finances is concerned, I would make the best of our lives. My children were not flashy or spoiled children, and I am grateful for that. I wouldn't know how to handle ungrateful children at all. I don't think I am equipped with that amount of patience. As long as they were clean and I was able to provide them with the things our budget could afford, they gave me no added grief. They had things that other children had such as electronics and cell phones, but I made it a point to teach them the value of a hard-earned dollar as well.

Normally during the Christmas holidays, since our schedules opened up a bit during that time, we remained consistent in our tradition of traveling versus decorating a Christmas tree and opening presents. We had learned to appreciate that time set aside for us to bond, catch up on the latest and irritate each other with our own personal nuisances and laugh at them later. We took many pictures to savor the memories as we grew older. I always would find time to sneak away as they slept, find a small shop that was still open, and buy them a little something to remind them that I had not forgotten it was still Christmas.

Once Christmas had passed, we decided to go to Los Angeles. Our intent was to go to Disneyland, enjoy the parks and of course see all the Disney themed lights and firework shows. While there we would spend some time with my high school sweetheart's sister and her children. He was still deployed and had reconnected with them after a distance of a few years. He knew they had faced some hard times in his absence so he wanted to try to make things more cheerful than they had been in previous years. I had no problem whatsoever helping him make this moment possible for all of them. It was made easy since they lived in Los Angeles as well. I could "kill two birds with one stone" was the plan I had in mind before I began to travel. He and I discussed a budget and I would purchase at least one thing that each of the kids really wanted and he would transfer half the money for whatever it was from his account into mine and I would cover the other half. We had recently set that capability up. His sister was overjoyed about this idea to the point of tears.

So, we made the trip and I prearranged for us to stay with my cousin who lived near Los Angeles as well. Once we arrived, she and I went shopping for everyone, including my kids. It was definitely going to be a great Christmas for all. The only request his sister had was that she be allowed to prepare the type of traditional southern meal she had grown up with as a little girl in Arkansas. Therefore, the next day after our arrival, I shopped and got all the ingredients for the meal and we filled her home with the aromas of the Deep South. Our children met and hit it off instantly. Everyone enjoyed their gifts that they received and the kids played in a backroom together as if they had known each other for many years. His sister and I shared our life's journeys over the simmering pots. We cried at times and encouraged each other that better was definitely in our futures although we both knew it was not going to be an easy task, but well-worth the sacrifices that we both would have to make.

We received calls from the war-torn countries of the Middle East as her brother called and wished us all a very Merry Christmas. My kids and I enjoyed our time so much in Los Angeles that we forgot all about Mickey and Minnie

Mouse and their display of lights. His sister was apologetic that we didn't continue on with our original plan, but as I assured her, Disneyland was not going anywhere at the amount of money they hauled in and we could always come back the following year. What made that Christmas magical for me was the ability to spend time with a family that appreciated us being a part of their lives and our investment in every single smile, laughter and even the tears of encouragement throughout the holiday. I gained an entire new extension for my family.

A few years had passed and as the saying goes, "Time waits for no man." I had witnessed soldiers I went to nursing school with advance and become Registered Nurses in their respective states and pushed on to higher degrees. I had helped soldiers, as they came back from basic training and AIT, not face the same challenges that I went through when I came home. I laid out a blueprint for them which outlined the hurdles they would face and helped them to avoid the extreme delays in getting back on their feet. I had a wealth of knowledge and resources regarding transferring licenses, references for jobs, transcripts, processing bonuses and advice for how to stabilize families while everything seemed to be unorganized and out of place. They knew they could call me at any hour and I would answer the call with no reluctance on my part. I have learned to be a soldier that was not afraid to lead and face challenges under fire.

I returned to my former employer in the administrative sector as I continued to apply for positions in nursing. However, it was not easy as most employers were only hiring internally or they were looking for persons with years of experience. This was not my original plan. I thought jumping into medicine was going to be an easy transition because I constantly heard that there was this shortage and demand for nurses worldwide. But it seemed more like a field that was not just based on skill, but on who you knew versus what you knew to get through the doors or even into an interview. I found out quickly that LVN's were being phased out as RN's, Medical Assistants and CNA's were being favored for positions. The process made it difficult for me, because the school I attended decided to take my military transcript and apply the units and time of achievements as elective credits only. So I was pretty much retaking many of the courses I had already completed in the military and now completing them again at the Community College so they could be recognized as credits in the state of California. This was beyond discouraging, but I refused to let it get the best of me or derail the plans I had set out for myself. Retaking courses had definitely set my timeframe backward for how long it would take to complete my degree. However, with all the set-backs and hurdles I faced, I eventually had the honor of crossing the stage and receiving my degree.

A year ago I witnessed my son receive his high school diploma and his acceptance to Dixie State College of Utah. My excitement was far more about his accomplishments than about my own as I recalled a young black male who not only received his diploma, but earned it despite the many challenges and temptations he faced throughout society. He also worked for a grocery store chain and played football for three years along the way. California implemented a special exit exam that would determine if a student would complete high school with a diploma if they passed it or if they would just be considered complete with just high school requirements if they were unsuccessful. So with my encouragement during the summer months along with his football practice schedule my son would attend classes to improve the success of him completing the required special exit exam. With that added assistance my son passed the first time he sat for the exam.

That was one less thing on his plate to worry about, as some students were working to pass this exam well into their senior year in order to receive a high school diploma.

There were many nights where he would still be awake long after his sister and I called it a night, and I'd wake up to a sliver of light shining through my bedroom door because he knew he still had to knock out a report, finish those mathematic equations, or tackle the not so entertaining subject of history. I took great pride in witnessing him not throw in the towel and push past the obstacles both academically as well as personally. He met some amazing friends as well that had similar drives and set their ambitions in high school. I took them under my wings as a surrogate mother figure to them, rooting them on in their accomplishments as well. At the high school he attended we parents built a small-knit group that was more like a family to keep our children encouraged and on the right track. They all grew up together and formed a brotherhood that would extend well beyond their adolescent years.

Before that emotional trip to send my son off to college, there was a request that I received by mail from Vandrick because phone calls were scarce at his location. It was a letter asking to see his son before he moved to

Utah to start college. He wanted to take the time to let him know how proud he was that he had not chosen the street life even though it was at his fingertips. And, he wanted to give him encouragement to make the best of this opportunity that he was being afforded. He knew from his own upbringing that not every parent would make the sacrifices I did to be able to have the finances to make college an option. He wanted to make it clear that he felt terrible that he had not been around like he needed to be in order to contribute to our son's successes that he had to witness through phone calls, photos and letters.

So, our son and I made an almost five hour, one-way trip to High Desert Prison in Susanville, California. Luckily it was summer when we went so there was no snow we would have had to face out on the road through places like Lake Tahoe and Reno, Nevada. It was a very long trip for me and one I needed to be well-rested for since some of those roads would be a bit nerve-wracking to travel if we were not alert. We hit the road as we had done numerous other times in the past. This visit was a little bit more special as we were not only celebrating the successes of our son, but it was also the weekend of Vandrick's birthday. We were able to visit and he would not spend his birthday in a cell, wondering if his family was thinking of him.

We made it safely to the facility and passed all of the security checks as we had for any other facility. We even did our normal routine of going into the volunteer's trailer house to freshen up a bit and made sure our clothing met the standards so we wouldn't have a need to go back and forth, changing our clothes that the CO at the desk deemed unsuitable for visiting. I never put any major effort into what I would wear into a prison because I knew that would be the day the staff was ready to pick apart my outfit, piece by piece. The bad part about it all was it seemed like they got some bit of joy in watching the visitors undergo the hassles. They knew when they were able to wear their civilian clothes off-duty they did not want to look less than favorable. However, that no longer mattered to them when they were scrutinizing a visiting family member. So I prepared for this visit by wearing simple jeans, a pair of sneakers and an Army jersey I got while I was on active duty. Our son was wearing a white Polo shirt and some black jeans, but he was given a heads-up by the volunteers that sometimes they don't allow men to visit in simple white shirts. So we might be sent back for him to change. We all looked confused at such ridiculousness, but we proceeded and prayed there won't be any issues today.

We were able to not agitate the COs at the in-processing desk with the white shirt and were approved to visit all day from about 08:30 to15:00. Once in the visiting room the first acknowledgement I received was for the jersey I was wearing as the Officer must have been in the military or knew someone who had as he asked, "Are the numbers on the back of your jersey your MOS?"

I responded with a simple, "Yes."

Then he wanted to know, "Which one?"

I wasn't sure if he was quizzing me to see if I was being honest or he was inquiring for his own personal knowledge or it was just his way to let me know he was affiliated with the military in some capacity. I replied. "68's are for the medical MOS's."

He gave me a nod of approval and then told us where we would be allowed to sit. His partner at the desk with him had little to nothing to say, watching us as we had our brief exchange of conversation.

As Vandrick entered the visiting room, I realized it had indeed been a few years since I last saw him. He had grown out his facial hair, although neat in appearance it was a definite change from the past. His facial features had matured a bit over the years, too. He gave us a smile of relief when he saw us and then grabbed his son who was much taller and bigger now, with that same ol' fatherly bear hug. I remained seated like I always did and waited until I was greeted because I felt none of these visits were ever about me. This time was allocated for father and son to reconnect, catch up and bond. I try to make the trips about our son and me providing the transportation as much as possible. I always made sure I kept Vandrick up-to-date through our methods of communication on things that transpired with our son over the years, so there was no real focus that needed to be on me.

Vandrick gave me my cue to stand up and he greeted me with a hug and one of those once-over looks from my head to my feet that men give women as he made an assessment of me. I knew he could see that I have not grown an inch in height since he last saw me. And, that I was no longer the super skinny person I used to be. I gave him the eye and grinned in acknowledgement of his assessment with a burst of laughter as I let him know much had changed about him as well as I greeted him with, "Hello, old man!" reminding him he was a grandfather now.

He took a deep breath in and exhaled, as he replied, "Yeah that one caught me off guard, but I accept that my kids are getting older and they are now having children themselves." He looked at our son to see his reaction to his statement about whether or not he was planning on having any kids soon. Vandrick and I both look at him and he looks back at us with a facial expression that confirmed he was far from ready to have any kids.

We all sat down and they began their visit. There was time for eating because these two very muscular men were not going to pass up a chance at food. They walked and talked as I looked around and noticed there were only four other inmates with visitors in our visiting room. The room across from us was far more crowded, but I am pleased that ours was not. I don't like congestion and in a prison that feeling was elevated times ten as I became more aware of my surroundings. I believe the military was responsible for heightening my body's alert system as well as contributing to my lack of sleep.

The photographer came and we took a few pictures. I didn't want to be in any of them because I felt I looked a mess, but Vandrick would not allow it. So, based on my height, I was placed strategically between two towering statues. We smiled and thankfully I was released from any more photographic moments, although they took a few more photos of just the two of them. Then we all went back to the elementary sized chairs and table that rested uncomfortably against our knees like it always did. We continued to discuss various topics. At some points the topics would be more emotional than others so there were moments when tears were wiped away and then the mood would change to a happier subject where our laughter would fill the almost empty room.

It wasn't made clear if we were expected to come back tomorrow and spend another day with Vandrick or if we were supposed to only visit him today. If we were to visit him tomorrow I needed to make reservations for a hotel. I would be willing to stay in Reno and drive back, but I was not willing to make that long trip all the way back to the Bay Area and then return. In his letter he stated he wanted us to come for the weekend, but I guess that had changed over time because he told me, "Tomorrow, my wife is coming to visit." As I was looking at his face he didn't seem happy that she was coming. I would normally shy away from the subject, but since he'd brought it up and he didn't look pleased, I forged on with the subject.

"What happened?" I asked curiously.

"Since she knows that you and our son made the drive to see me today, she told me she was now coming tomorrow," he shared with me.

I tried to mellow out his tone with a simple, "Well, it is your birthday weekend. I can understand it." My attempt at calming his agitation did nothing.

He quickly replied, "It has nothing to do with my birthday! It has everything to do with you being here!"

"Oh, okay," I responded and after a brief silence allowed the conversation between father and son to continue. I tried to act like I didn't already know his wife had some underlying personal issues, with one of them being me; but I was a bit shocked that they were having such a deep effect on Vandrick and it was now visible. I didn't want to discuss anything further as it was not the time or place to address it. Our son going to college and their time together was the most important thing to me.

The two of them discussed football and the injury our son was recovering from because he tore his ACL in the second to last game of the season. Therefore, he was going to have to postpone playing football in college. He needed the downtime after surgery to heal properly. He was cleared by his surgeon but he still wanted to gain as much strength as he could before he got back into football. So, the first semester of school his plan was to acclimate to college and take classes that would familiarize him as much as possible with the expectations of college versus jumping right into sports. They laughed and talked about the sport for a while as Vandrick heard the aspirations of our son to not leave football behind. He just wanted to give himself time to be able to fully perform at his best. Vandrick told him he would be waiting to one day watch him playing on his TV in his cell. He would let the brothers know to look out for his son, too.

Vandrick shared many of his own personal aspirations, goals and change of heart, mind and spirit over these years. His perspective about things from the past until now, were so vastly different than when he first started doing time behind bars. Things such as the effect of losing his mother and not knowing where his brother was; how he realized even though he did not commit the crime that he was incarcerated for, it has given him an opportunity to change and accept the fact that the decisions and choices he was making out in the streets were heading down a path of either prison or death. So, despite all the anger he had about his situation he was grateful that God had a calling on his life to save him, even if it was through prison.

He still spoke of wanting his freedom, but he spoke of it differently now. He spoke of it from a place of peace. He shared with us that if he was granted an early release it would be based on God's will and not his own. He finally told us about the many times he had legally tried to get himself out of prison and each time it had ended in failure. Or, how he thought people who cared and said they loved him and were supportive of his destructive behavior had vanished and were not a part in helping him in this new state he was in. Once he decided to stop being self-destructive and participating in negative behaviors, that was when he saw all the people who used to surround him, his inner-circle, disappear. It was a bittersweet moment as he spoke candidly about his feelings, but we listened to him attentively. He even broke up the somber moment with what I took as a joke and as usual I disregarded it as he said, "It'll be funny when you and I are together after all of this is over."

I laughed and our son looked at me. I just rolled my eyes at both of them and replied as firmly as I could, "I know you are Muslim, but I do not believe in polygamy, so cut it out! You are already married!"

In his normal tone and with a sly grin, he responded without hesitation, "That won't be the case by the time I'm out of here. I've said this to you before."

I recalled him mentioning this to me before but I never took things like this that he would say seriously because I had always known of Vandrick and his player ways. There had always been multiple women at a time in his relationships and his ability to pull in a few more. So there had never been a need to put stock in any comments he made about his plans down the road regarding women.

He added, "Just you wait and see. When we're about 50 years old, you and I will be married."

I laughed at his foolishness yet again. I reminded him that would be torture for the both of us. He wouldn't know what to do with me and I wouldn't know what to do with him. He was celebrating his birthday and mine had just passed four days prior to his. Our zodiac signs would clash because they were the same—Leo. Those two lions would fight until the bitter end! He assured me that would not be the case because he knew me better than anyone else. He told me he knew exactly how to take care of me. I was the female version of him. I said nothing else to stop further comments.

We changed the subject to a lighter topic as we realized the visit was coming to an end. For the first time this visit didn't seem to be one that dragged on for hours. It honestly felt extremely short. I guess because it had been such a long amount of time since we were all last together, we had plenty to keep us occupied and engaged and the time just got away from all of us. Soon, they were saying goodbye with bear hugs and Vandrick's well-wishes to his son as he goes off to college, reminding him that he would be waiting to find out the TV schedule for his future football games; promising him that he would be rooting for his academic and sports success from behind these concrete walls and barbed wiring; and a final reminder not to blow this opportunity that many were not fortunate enough to receive. He eventually turned to me for a farewell hug that overwhelmed my size yet again, and he thanked me for bringing our son to see him before I hit the road to get him off to his college campus. He also thanked me for doing a damn good job of raising him through whatever channels I chose to take. He let me know he would call and check on us as we traveled back home therefore, make sure my cellphone was charged and on. He continued to lecture as he told me if I got tired to stop at a hotel and drive on after I am well rested. He wanted me to know that even though our son was leaving he would still like for him and I to remain in communication and for me to schedule a few periodic visits to remain in contact. I took in all that he had to say, well almost all of it, not that part about "periodic visits." I had never visited Vandrick without our son and this was not the time to start making that type of arrangement or granting his request.

He released me from our embrace and said a final goodbye then our son and I were processing out of the facility, relieved that we made it but ready to head back on this five hour drive until we are finally home. Times had definitely changed. This was the first time ever my son could offer to take the wheel if I needed him to because he was a licensed driver who had a car of his own and had been driving for about 6 months now. But I knew I was not taking him up on the offer. I was still getting used to him driving regularly and I knew the strain this drive would place on someone of my experience, so it would only become a greater challenge for him. We stopped at the visitor's trailer, freshened up and headed off of the facility's grounds. We gassed up a few miles away then it was back onto the highway. This time I would need to stop at my mom's house because that was where my daughter was waiting for our return. Plus, I could rest up a bit there before driving the remainder of the way home. Her home was only about four hours away from the facility. After the long day we'd had, I needed the break.

We received the call I expected to make sure we were doing okay. Vandrick knew we were alright and told me he appreciated the time that we got to spend with him because a week from now a group of us would be taking the road trip to get our son safely off to college. My son was able to take his jeep with him so he and my daughter's father would be in his car with all of his things he was moving with him. I had rented a minivan that would be filled with our luggage and loaded with passengers: my grandmother, my mother, my daughter and my nephew with all of us entertaining ourselves and taking in the scenery until we reached Utah. My mother, daughter, and I had made this drive before when we went to tour the college and the area. My son decided he loved this school and knew he wanted to make this his college. It had a similar feel to the high school he had just graduated from. He met a few students that were going to attend the college during an all-night freshmen orientation and meet-and-greet in the gym. I even exchanged numbers with one of the student's fathers because he knew of an apartment complex not too far from the school that had all the amenities necessary to make our boys feel right at home. Plus, many of the football players were sure to be staying in these apartments versus the on-campus housing units, but they would still have the same amount of supervision.

My son's enthusiasm and eagerness for this college experience was definitely rubbing off on me. Even though I knew I would miss him terribly, I was excited to see him pursuing and setting his own goals and was something I hoped and prayed for from the day of his birth. We went, we drove, we saw and we all cried as we said farewell to my son and left him, knowing he could call anyone of us if he ever felt a moment of homesickness or doubt. We also knew he could call during the awesome moments although we weren't right around the corner to share in the joys. We wanted him to know we were all with him even with the distance. This was now his moment to get up on the saddle, take the reins of his horse and ride off into the sunset.

We made it safely home after experiencing my very first child's process of matriculation; another bittersweet moment in the books. There was not a dry eye in the mini-van as we drove away. We are proud of his accomplishments but sad to have to part ways until the next scheduled college break. We hoped he would at least, initially, want to visit us periodically before all the routines of college, his acquaintances and social clubs become the priority. As for my daughter and I, it was back to our normal routines of school drop-offs and pick-ups now that she was in middle-school. I still remain employed in the field of administration, in the military, and I am taking college classes of my own in the evenings at least three times a week which doesn't allow for much of a social life for me. I have taken on a higher rank with my unit and I'm required to travel more than I hoped. Thankfully, I had planned a quick getaway with my daughter to the Bahamas after her and my son's graduations. I got the feel of how tough and expensive it can become, accommodating two graduates in the same year—one high school graduate and one elementary school graduate.

However, a few months before the trip took place my son applied for a second job with his free time after football ended, so it was going to be impossible for him to travel, even though it was a trip planned especially for him.

But my daughter was adamant that I not cancel it completely so that she could celebrate her graduation. I didn't want to break her heart and I also wanted to enjoy the vacation as well, so we rearranged the trip around the two of us and bought my son graduation gifts instead. He seemed okay with our new arrangements because he got the gifts he wanted out of the deal which was an upgrade to the latest version of the iPhone and a Jeep that my mother decided was a necessary essential for his graduating year and transition into college. Based on his hard work and grades she felt he deserved such a luxury. I made sure to inform my son that even with good grades and a good work ethic I didn't get a car. I had purchased all my cars myself from the first to the one I currently drove. No one had ever bought me a vehicle so he'd better be careful and extremely appreciative of the purchase made on his behalf. There was a definite difference between being a mom and being a grandmother, just as the level of disciplining children changed for some odd reason as well. Grandmothers suddenly decided talking and listening was the new style of parenting while we definitely received an old-fashioned butt whooping and was grounded.

My old high school sweetheart and I remained in communication when we could, but his military career had him stationed at Fort Wainwright, Alaska. He was in the field pretty consistently so I had only been able to make a few trips to see him once he redeployed back to the states from Afghanistan. I did make it a top priority, taking vacation leave from both jobs and school to be there for his redeployment ceremony, and I stayed with him in Alaska for two and a half weeks afterwards so we could get reacquainted. That had been the longest amount of time we had been able to spend with each other over any given period. After I came back from a military mission assignment in Central America, I visited another time for about a week, based on his immediate request to spend some time with me. And, I was also able to do a few quick visits with him in other states as he traveled on orders.

We had an open relationship and were pretty stable in our understanding and the rules we had set, especially under the restrictions caused by our locations and job requirements. We had to deal with seeing each other, at most, about four times in any given year. The distance could put an extreme strain on any relationship and we were no exception to the rule as there were times we might not speak to each other and not even realize it until weeks had passed; and as history had proven, "when the cats are away the mice will play." We didn't want to add stress to our already fragile relationship so we wanted to eliminate the core issues of trust and jealousy by imposing a clause that we could date other people without penalty. Although the relationship was indeed open I didn't take up the opportunities and advances offered from male associates like I could have. It was primarily based on the fact that my commitment to excel in college took up a great amount of time and priority in my life.

However, my old high school sweetheart decided to engage more than I did. His schedule was less demanding than mine during this time. His rank in the military, responsibilities and his retirement years were just a blink away for him. He had started having thoughts of moving to a new duty station and what life would be like outside of the military after 20 years of service. So he started to entertain himself more than expected with a couple of young ladies that I was aware of. And over time one of the ladies became more established in a relationship with him because she was far more accessible to him than I ever could. So our communication became more distant than it already had been.

I appreciated the time I did make to spend with my friends. Lunch time with a co-worker or a rare evening dinner to catch up with the girls was the only socializing I participated in. That was my time to decompress from the new hectic rat-race that had become my normal. I was occupied with wanting to make sure outside of my two demanding careers I spent as much time as I possibly could with my daughter and kept up with her schedule. She had always been the child who took charge, knew what she wanted and would go after it. She was the same when volunteering for activities that she wanted to participate in. She knew either me or her dad would somehow make it to the evening performances, meetings, birthday parties with classmates, or would pay for whatever adventure she was heading out on; even if I had to sign the permission slip in my sleep.

My daughter was a teenager now and was very ambitious. She loved to engage herself in multiple activities and I had no reason to slow down her stride because she was very wise for her age, still a very youthful spirit, yet exceeded the expectations in school academics. Her father and I playfully argue often about who she was more alike. I think she is more like him with his creativity and with knowledge sharp enough to be a contestant on Jeopardy, but he begs to differ that she is very much like me with her ambitions, outspokenness and take-charge attitude. I guess we could agree she had a lot of both of us in her. Now, we're a pleasant mix because we have learned to co-parent better together than we had in the past in our attempts to encourage her growth and development. It took some time and a few intense hurdles, but we had gotten through them and we now respect each other's differences.

Vandrick's prison points had lowered and he had called to let me know he was moving to a prison he had been at before. He was heading over to CSP-Solano, Vacaville, California. He was content with this move because it allowed him to be closer to his family and he was hopeful it would re-establish that family connection and lessen the burden of that long distance drive. During this call, he said to me, "I will be calling you more once I get to where I am going."

"Ummm, why would you do that?" I replied quickly.

He responded, "Because there are some things I want to talk to you about and there I will have more access to call."

I said, "I am sure it's not necessary." Then I thought to myself, *"What the heck does he need to discuss with me?"* But I decided to leave that question in my head.

He asked if I had talked to our son lately and I let him know that he was doing alright and had even sent a few pictures while in class and he seemed to be adjusting well. Vandrick told me he planned to call and talk with him more often as well once he got to the new facility. I was pleased with that. I felt with the both of us keeping him supported as best we could he should feel encouraged to stay focused. The call came to a pretty quick end as the call dropped mid-sentence. Something must've happened up there. This is common in Level 4 facilities. I'll have to wait until I hear from him again to find out if it was an assault or a staff member asserting his authority and seeing if he could get a response out of the inmate by intentionally cutting off the phone call. Yes, that had happened to us in the past. A few days later when he called me back, he told me things like that happened and an inmate trying to stay on the right path had no recourse of action. So they just had to accept that some egotistical Officer wanted to throw around his authority and see if you would retaliate against them so they could place you in solitary. Moments like this were the times when you prayed a little bit harder that God would allow something to give and they would be able to make it out of that place in one piece and eventually come home.

Having a better understanding of the case that Vandrick was facing only further inspired the feeling that soon he would be able to find an attorney that could assist him with coming home. Over time, he had enlightened me in letters and phone calls that he was doing this time for the friend that came with him to our son's graduation years ago. And, the reason they didn't stick the charges to his friend and the other person involved was because Vandrick was being loyal to the rules of the street that you don't snitch on the people you run the streets with. He told me that when he didn't turn into a rat, the police officers assured him they were going to charge him with all the counts associated with the attempted murder. It initially started out as a scare tactic to see if he would break under the pressure, but quickly became his ultimate reality. He believed the evidence would at least force them to dismiss the charges and drop the case during trial.

However, he didn't realize how the system could either work for you or against you dependent upon who you were and how much you were willing to give up; he didn't expect the sudden death of his mother and how he had to put money he didn't have toward her funeral expenses versus securing his attorney; and of course he didn't expect the number of people that stole from him and tried to cover it up anytime he'd ask about money he knew he still had. Therefore, he ended up with what he referred to as a "Public Pretender" who lost his case. He did recoup some of his losses through his own efforts and schemes and filed the appeal to see if they would help him out and get him released, but he ran into several hiccups along that path as well: like hiring an appeals attorney who was disbarred and ran off without returning any of his clients' funds; or contacting the Innocence Project in Northern California and requested they take a look at his case because it was a case that had DNA evidence that could clear him, but due to their overload of cases he was given a response of, "If you can get the victim to recant his statement we can work with you. If not, then don't contact us we'll contact you," in so many words.

That was when he said he lost all hope that it was ever possible he was going to go home after the numerous failed attempts to gain his freedom. He was able to see people closest to him moving on with their lives and slowly forgetting he even existed behind those gates unless he reached out to them, where he always heard various versions of, "Aw, I was just thinking about you." He tried not to build up animosity against the things he saw and heard. Along with that and a few health issues, I guess your heart could only take so much before it started to show signs of wear and tear as well. So, he decided his best bet was to sit back and become stronger in his faith as his mother told him in her final letter, "Give your life over to God". He did that and became stronger in his new found identity of Hamza "the Lion," the Muslim and relied solely on his Higher Source to make a way out of no way.

CHAPTER TWELVE

Listen and Learn Something

I was sitting in my cubicle one day in September and I received that familiar call from the prison which announced, "This is a collect call from Hamza, an inmate at California State Prison Solano." I accept the charges as I hear on the other end, "*As Salaam Alaikum.*"

I replied modestly, "*Wa Alaikum Salam.*"

Throughout the last year or so this had become the greeting when he called. He had extended the teaching to me and even to my daughter. We thought we were to reply "*Laikum Asalam.*" He corrected us and we practiced to make sure we didn't get it wrong thereafter. He also wanted us to know and understand that what we were saying was just a kind greeting of peace and blessings being given from the greeter to the recipient and then returned back to the greeter. He was making it a point to educate as much as possible and to debunk many of the misconceptions that were taught. And he knew I never turned down an opportunity to learn whenever possible. It irritated my soul when we have access to so much knowledge yet people walked around ignorant and would even shut themselves off from learning. After we exchanged greetings he asked if I had listened to the phone announcement. He was no longer in Susanville and had made it safely to Solano.

I responded, “Yes, I heard. How do you feel after the move?”

He let me know he was extremely relieved to be away from that other prison because there was a lot of racism and discrimination festering up there and it came in from everywhere. That explained our last call when the call abruptly ended. It was confirmed it had been an Officer that hung up in my face. Although he had been provoked to react, he chose to lockup in his cell and read the Qur’an and recite more Arabic chapters and verses to calm his nerves as well as pass the time. He knew getting to a lower level was more important than fighting an Officer that was already set to win, even if Hamza could physically defeat him. The system would win with their actions against him. He showed he was learning something from all of the time he had done because the inmate, Boo, who first entered the gates of prison would have never thought out the long-term consequences and would have been in the “hole” or shipped off to a prison even further away. Well, he didn’t want the call to end before 15 minutes so he quickly reminded me of our conversation.

"I will be contacting you more often," he stated.

"Ummm, that's not necessary," I reminded him again.

He laughed and said, "Yeah, okay," letting me know he was not taking what I said seriously at all.

We heard the 60 second warning that the call was about to end. He gave me the greeting again and I likewise did the same as the call disconnected. But, within five minutes my phone was ringing again! After accepting the call and after we greeted each other, I asked, "Are you for real?"

He laughed and said, "I warned you already so you might as well get used to it."

I replied quickly, "Aww, hell nah!"

He laughed and continued to talk as if I was not sitting at my desk and baffled as to why he had so much time to call, and why was he calling me! Luckily for him, I was able to accept his call and entertain his craziness because I could use my headset on my phone and talk to him while I worked on my cases at the same time. We engaged in small talk and caught up on things since it has been an extended period of time for us to really catch up. He asked a couple of questions about what I had going on and what my priorities were. I assured him I was involved as I had been over the years with children, work and school and pretty much in that order.

He asked me a few questions related to my religious practices and I let him know that I was still an avid Christian although I was less able to attend church regularly like I used to in the past. However, as time permitted, I did attend a few services at the popular church I had always attended. He even got a little more detailed by asking about my practices of fasting and prayer. I informed him that I did both and based upon my teachings, I shared with him the hours in which it was done. He shared with me that he fasted and prayed from sunrise to sunset which I believed I could do. I normally broke any fast at 3:00 p.m. We weren't able to talk much longer because the 60 second warning had signaled us again. However, he told me he was going to take care of something quickly and would call me right back. I let him know it was close to my lunch time and I would be available. I normally took lunch with a co-worker that he is familiar with because he lived in the same area as my mom's old house. He asked that I give a "What's up?" to him and he would be calling again.

Just as he'd told me he did call back. This time my co-worker and I were in his car, going back and forth about what we were going to have for lunch. I never knew what I wanted and he should know this by now, but I kind of wanted a sandwich so we were heading to a deli in Oakland Hills to get one. I answered the incoming call from Hamza and accepted his call, yet again. One thing I was thinking was how grateful I was that over the years the collect calls had gotten much cheaper than they used to be. We talked a bit and then the fellas caught up a bit as well. They hadn't talked in over a decade so it was a bit of catching up for them as well. Reminiscing over past friends, learning some were doing well and some were not, some locked up and, unfortunately, some had lost their lives. Neither Hamza nor my co-worker were still committed to keeping up with the streets and what little they had to offer. My co-worker had not only grown past that former life but was married with two children and graduated with his Master's degree. He accomplished this during the same time I was away, completing Basic Training and AIT while living a life that was not expected of a person who grew up in the environment we all did. He had definitely defied the odds.

It was unfortunate Hamza hadn't followed a similar path, but he was fulfilling his goals and making accomplishments as he worked on his growth even in his restrictive circumstances. He proudly shared with us that he had completed his GED; was working on his college degree; was enrolled in lots of courses as well as mentoring other inmates who were trying to get back on the right path. Soon the call was coming to another end and we said our greetings as the call cut off. He had assured me it was not his final call because he still hadn't told me what he had been waiting to tell me. He said he would give me time to enjoy my lunch and get through my day before he called me again. I was curious to find out what he had to say, but it would have to wait. My co-worker and I ordered our sandwiches, chips and drinks to go.

We returned back to work and the day continued without a hitch. I maintained my normal routine of classes that evening. The only change in events was when I received a call from an already expected caller, Hamza. I was shocked by his ability to call as often as he said he would from now on. I was now wondering was he the only person that had access to the phones? I would think other people would want to use the phone, but I really didn't know how it was set up inside so maybe there was a wall of phones so the line was not that long. We greeted each other and he asked what I was doing? I told him I was at school and I take science classes so about this time was when we took our break between lectures and labs. His timing was on point today. I considered him as being lucky because my day never worked out like this. He made a sly comment that it was because of him we were having such great timing. As usual, Hamza and his cockiness never failed.

I could only say, "Yeah, right" to his moment of patting himself on the back as I laughed because some things would always remain the same. I became serious and asked him, "When do you plan to tell me what you've been saying you want to tell me?"

"Not right now, I am planning to call back again," he replied. He assured me I need not become occupied with it and allow it to distract me from my work and school. I needed to concentrate on those things.

This time I had to shorten the call because it was time for my lab and I needed to get in there and help my group out. So, we ended the call with greetings and in no time I was back with my group, determining wavelengths. It had been a long day. I picked up my daughter from my grandmother's house and headed home to prepare us for the next day. This was a part of my normal routine and of course the days were long and the nights were short, but before I could call it a night I had homework that must be done. Luckily, my daughter was consistent in getting her homework done and my grandmother guaranteed me that before she could go out and play with her friends her homework came first so when we got home she wouldn't have much to do aside from figuring out what she'd wear the next day to school and take her bath before bed. My grandmother would have fed her dinner as well, but if she was hungry she would occasionally get a quick bite to eat. But oftentimes she was tired so food was not on her mind at all, especially since we are up and out early in the mornings. I must get her back to my grandmother's house so that she could walk to school with her friends and I could get to work on time.

We have had this discussion more than once about her staying the night with her grandmother instead of coming right back in a few hours. But, she loved being up under me so I didn't push the issue. She would even sleep with me with no argument if I needed to keep the light on most of the night until the early morning hours as I sit in the bed with my computer on my lap, typing away. She would tolerate the noise and lights just to be next to me as she slept as if nothing was going on around her. I generally get about two to four hours of sleep a night and I was used to it. I run throughout the day as if I have slept all night, but that is surely not the case. The people that knew my schedule did not understand how I was able to maintain it and I couldn't explain it so it was not much of a subject just occasional comments of, "You don't sleep at all," or they just shook their heads anytime they thought about it.

I was back into my routine and my phone rang. I tried to determine if I was going to break away from my studies to take the call. After a couple of seconds I decided I would take it and it was Hamza, just as he promised. I can't put my finger on it, but this time he sounds a bit different. We greeted each other as we always do and he sighed and went into what I assumed was the long-awaited conversation. He starts with, "Ever since you left from visiting with our son I have been trying to work through some things. I haven't been able to get you off my mind…"

This threw me for a complete loop because Boo, Vandrick, or Hamza had never been like this with me. I didn't interrupt as I listened and allowed him to say everything he needed to say.

He continued, "… there were some things that I had thought about before that time, but I had never approached you because I wasn't ready. I didn't want to come into your life and have things not be right. You are an amazing woman who has done an outstanding job raising our son to the best of your ability. I always commend you on the sacrifices you have made to provide for him. Since I last saw you I have been fasting and praying as I regularly do, but I have been seeking guidance specifically about why Allah, which means God in Arabic, kept sending me visions of you. When I try to shy away from it He reveals a sign that you are who I am supposed to be married to. I couldn't tell you until I was certain and Allah revealed it to me. This is why I was waiting before saying anything to you. I had some things that I needed to take care of and in time I will complete those things that remain on my plate. Therefore, I am asking you to trust me that everything I am saying to you I really mean. Just be patient with me through this."

I was at a complete loss for words. There was nothing I could say after listening, but I heard every word that he had said. I was in shock!

He continued with, "I know this is unexpected and a lot to share with you. I am prepared that if I cannot have you as my wife—I am not requesting you as a girlfriend because that is not permissible in my religion and I want you to be clear that I am serious—then I am fully prepared to finish out the end of my sentence BY MYSELF! I am certain about what I am willing to accept and what I do not want."

He had covered many of the things that I would have questions on, but I still had to leave no stone unturned. So, with a bit of reluctance I asked, "So what about your wife?"

He stated, "I've been in the process of taking care of that. It will take some time."

It wasn't the fully detailed answer I was looking for, but I was able to hear that he was taking that major conflict into consideration. I continued by asking what seemed like a question filled with shock and disbelief, "Why me?"

He replied, "It's been you for a long time, but when you were leaving to join the military I was going through a rough time. I was still filled with foolishness and focused on getting enough money to get an attorney. I hadn't fully stopped doing things that could jeopardize me. I was angry at the situation I was in and wasn't in the right mind frame for you. You were committed towards taking care of what you needed to do to support and provide for your family and I was in a very selfish state-of-mind at that time so I wouldn't have wanted you to leave. I tried to tell you back then but I couldn't go all the way through with it because I didn't want to hurt you. So, instead of stopping you from the goals you had set out to complete, I turned to what was available here in California. I was hurt from the divorce I had just gone through at that time, but I was not going to stop her from the decisions she wanted to make, so I gave her what she requested from me."

My thoughts were swirling around me. Over the years I had always heard him speak about his first wife in nothing but a positive context so I asked a more suitable question that clarified my other question. "So why not go back to what you had before, especially since you've spoken so highly of that situation?"

He responded, "I can't because she left me for dead back here. She cut off communication with my children which meant I lost the relationship I had with them as well.

That hurt me the most. My kids and I had a bond that was connected and it was broken. It took a long time for me to heal from that hurt and because there are those things and a few other things you don't know that I am not ready to share with you at this time, I won't ever revisit that relationship again."

I did not expect to hear this, but to be honest I was not prepared for any of this. But, since the conversation was on the table I was not going to shy away from it. Well, I might as well pay attention and focus on what was being said.

"...also, once I am cleared to have visitors come I would like for you to come and sit down, just the two of us, so we can talk face-to-face. I don't want this to just be a telephone conversation that you hear and not get the chance to look me in my face. I owe you that. I wanted to say all this to you in person, but I have to go through clearance first with my counselor before they will allow contact visits. It may take a few weeks, so I couldn't string you along with all the calls and you not know my intentions. Are you willing to come and visit now that you know my feelings?"

After a brief, yet necessary, pause I replied, "Yes, I'll come and visit."

We laughed and talked about the past and he made a few crazy jokes that I couldn't laugh too loud at because my daughter was still sleeping right next to me. He could definitely be silly when he wanted to be. But he also had turned into an old man over the years and had a set bedtime. He jokingly told me he had to get his "beauty sleep" every night but he made up for it by waking early to perform his first morning prayer. He taught me that prayer in Arabic was "*Salat*" and the *salat* was named Al-Fajr. Our call came to an end with the greeting and I was sure with the day and evening I'd had I needed to rest, but I couldn't. That conversation was running on a loop through my head. I was really in a state of shock!

Hamza had opened up to me in a way I had NEVER heard him speak before. If for nothing else at this point I respected him being candid about some things, but I was still asking myself, "*What in the heck had just happened?" You have got to be kidding me? This is crazy, right? Is it not crazy?*" We still had a lot of communication and layers to get through, but in my private thoughts I was at a point in my life where—I am certain because I prayed to God and promised with all sincerity just a few months ago—I would not put on another set of track shoes to chase after, persuade, or force any man to be what I desired him to be. He would be what God had designed him to be and he would complement my life and I would complement his as well. No more build-a-bear relationships in my life. I was fully content if it was in God's will for me to be by myself. I had also prayed in that same prayer, "God, if it is meant for me to walk the rest of my years alone then strengthen my heart and my spine to endure as You see fit. If I am going to trust in Your plan I will trust it fully and completely."

CHAPTER THIRTEEN
K.I.S.S.

Hamza and I had spent some valuable time on the phone, communicating like high school love interests. It was cute and refreshing but definitely still a shock to me and my guard was still pretty high. I was taking every moment slowly. It was a tactic to protect my heart, even though our conversations were fulfilling on so many levels. But, this early in his courting me I was not willing to give him all of my trust. There were still so many layers we had to get through. The most important one was he was still legally married to his second wife. No matter what he was saying to me verbally, I would not be able to move confidently in any direction with him until I was certain that this was not a game or ploy that I was participating in. I could hear every day that this was not the case, but actions always spoke louder than words.

A portion of our communication I appreciated was his request that we fast and pray together. This was new for me because I had always done this on my own, intimately within my relationship with God. But, there had been no relationship I have ever had with any man where I'd been asked to join him on his journey to connect with

God. This was a first and I respected this aspect of closeness and the bond that we were building as well as the spiritual covering he was trying to place us under. His dedication to Islam was definitely reflected in his daily life as he constantly demonstrated he had made a solid commitment to become a better person and give his efforts towards doing the right things, even if they were difficult.

I complimented him on something else in a conversation we were having. He asked, "What is your perception of me at this point and through the few weeks we have spent with each other?"

The one thing I could recognize and commend him on was his consistency! That had truly been a quality that I had come to appreciate about him at that time. He remained constant in his words. If he said he was going to do something, almost 100 percent of the time he followed through. With the flakiness I had seen in my past relationships this was definitely a refreshing character trait that I admired.

My guard was still high when he was cleared to receive visits and the time had come for me to make the drive to see him. I had butterflies in my stomach and my nerves felt like they were all firing off at the same time. Even though we had a grown son and I had known this man ever since we were teenagers, it did not in any way remove the nervousness I felt at seeing him from a different perspective in my life. Not to mention that when we were younger and engaged in all that sex we were having, we never crossed the boundaries of being nothing more than friends with benefits. We had just straight sex—"hit it and quit it" was a popular term during that time—so we experienced no chemistry or intimacy. As adults, this would be the first time Hamza and I would share a romantic kiss and the thought was freaking me out! What would this be like? He had always had a way of turning me on and our chemistry in the bedroom was undeniable, but this was next level stuff for me, with him! Romance, affection and anticipation tapped into all those emotions I had never experienced with Hamza (Vandrick) until now.

I hoped I hadn't dressed up too much. I had put far more effort into my appearance than I had ever done before. I knew a safe color was black and it wouldn't have me going back and forth to my car to change. I had shopped for a pair of black jeans and a black top to pair with my open-toed bootie heels so I wouldn't appear so short next to his big stature. I have a short haircut which I kept pretty much after my first year of active duty. I got tired of putting my hair in a bun, so it was easier to keep this style to avoid any issues with regulations. I also grabbed one of my son's black and white striped sweaters he no longer liked just in case it was cool in the visiting room. I didn't think I would need it outside because it got pretty hot where he was.

I had never been to this prison to see him even though this was his second time being there. I had to acclimate to this environment because for some reason everywhere you went within the system everything was different. Some facilities followed the rules to the code and others seemed like they were making the rules up as they went along. I didn't know what I was in store for so I had packed an extra bit of patience along with me, just in case I had to go into my reserve tank and use it because some of these staff members could be outright rude and some could be straight up stupid. However, I gave them all a fair chance to prove to me what their attitudes were then I responded accordingly. For the most part, I had encountered a pretty fair and cordial staff. I was not looking for special privileges, a hook-up, or was planning to make any friends, but I did expect professionals to act and remain professional.

I was making a last-minute trip because Hamza unexpectedly found out he was cleared to have contact visits before his actual meeting with his counselor. He called me as soon as he was aware and asked if I could come to see him. There was a system I could have used to make an appointment. However, due to the short timeframe I decided to show up and do a walk-in visit instead. I had gotten my pass and waited for my group number to be called, just as my son and I had to do over at Salinas Valley. I made it in to be processed and I was watching everything. I was familiar with most of the processing but because things might change from facility to facility, I didn't want to be singled out and embarrassed. So it was easier to just pay attention.

They had two lady Officers at the desk and clearly one liked her job and the other one was there for the paycheck. The one who seemed to take pride in her job was pleasant, spoke civilly to the visitors and even engaged with the children as they approached the desk. Some of the kids were restless and groggy. I definitely remembered those days from the past. The first lady Officer was giving those kids a bit more of her attention as she joked with them so she could get the information that she needed for them to disclose to make sure they were the child listed on the birth certificate that their parent or chaperone had provided before they could come in. However, the other lady Officer was sucking the life out of me so I could only imagine how the kids were feeling as she started barking orders at them, trying to extract the same exact information. I guess I would blame it on the pregnancy, since I could clearly see her belly was protruding underneath her uniform shirt. A difference in attitude goes a long way because I could see some people were choosing to allow other visitors to go ahead of them in line to avoid her less than desirable personality.

I was not bothered one bit and my son was not with me so I didn't mind she was the one who had to check me in. I knew how to blow past an unprofessional person very easily by keeping my answers short and sweet. She was going the extra mile to not make eye contact with anyone so I didn't have to engage her with a fake smile or pleasantries. I just needed to answer the question of, "What line did you sign-in on?" and I would be done. I did make it a point to say as she handed me my pass and military identification back, "Thank you and have a nice day!" I hoped it would help her realize she was in a customer-based business. It must have caught her off guard because she did look over her computer screen for a glimpse at me and gave me a small grin. I imagined she hadn't done that in a while. I was thinking as I proceeded to the machine to get tokens for pictures and then walked through the metal detectors. I had no issues there and I headed through the sally-port of fencing lined with barbed-wiring around the top and downstairs to where I would give the Officer in a small office my identification and pass and they would notify Hamza, back in his housing unit, that I was there and waiting.

I am overwhelmed because this was the largest visiting room I had ever been in at any facility I had ever been at. This was definitely an intimidating environment and I was less nervous about seeing him than I was more concerned about the number of people in this room. The place was the size of an elementary school cafeteria. I was not about to roam around this room to find a seat. I took the closest one I could find. Thankfully it was right next to the grill-gate that I had to enter in order to get into this visiting area. The one thing that was standard no matter where I went or what facility I was in, they definitely shopped in the same places for their furniture. I am back into these kindergarten chairs placed in front of the knee-high kindergarten tables. I am facing where I can see the entire room and only my back was to the grill-gate so I felt more comfortable.

I was glad it was not long before I looked over to my left where the doors were through which the inmates would come in and leave, and surely I saw Hamza in a white kufi with beads around and streaming from his neck, smiling as he approached me. Our eyes never broke from each other as he made the walk in my direction. I remained seated until he got closer. I stood without being prompted this time and he greeted me with a slightly different bear hug this time. His hands traveled a little bit lower and planted themselves on my butt as he pulled me in. I was glad I was wearing heels this time which made me a little bit taller. I was at his shoulder for the first time ever and he whispered into my ear the greeting, "*As salaam Alaikum*."

I properly responded, "*Wa alaikum salaam*."

Then he turned his head ever so slightly and for the first time I was romantically kissed by my son's father, Vandrick, or as I now called him, Hamza. Every ounce of me was tingling, but that fire couldn't be put out in here. It felt good as we parted and looked at each other because we both knew we had never embraced like that before. We had confirmed through that look that we were satisfied with what we had started. We sat down with him sitting on the outside and me on the inside. He placed his hand on top of mine and placed it on his leg. He thanked me for coming and asked questions about processing in. I told him it wasn't bad and the waiting to come in wasn't too long, but that in the future I would prefer an appointment so I wouldn't have to arrive early and wait until my number was called. The people with appointments had priority and processed quickly. As we continued to talk, I listened attentively to the things he shared with me and in return I shared things with him. We spoke about our intentions towards one another. He still had obligations to his soon-to-be ex-wife and was working through all of that. He tried to keep me out of that part of his life as much as possible because he realized I didn't want to be involved in that. I preferred to leave that up to him. He had known for years my feelings regarding her.

We had always tried to be honest with each other when it came to our relationships because those were people that would be connected to our son in some way. I knew his feelings about guys I had dated and he knew mine about his wife. The manner in which she carried herself and the lack of interest she took in his children would never sit well with me. There was an old saying that stated, "A woman with children was a packaged deal." Well, the same remained true for a woman who chose to be married to a man with children. He was "a packaged deal." It was even more of a responsibility to put forth the effort to be involved with the children on behalf of your husband, especially if he was incarcerated. She was the one who decided she wanted to get married to him while he was locked up, so it was not like she walked into this situation blindly. I could only imagine the amount of hell he was taking on his end, attempting to dissolve his marriage. I'm sure he was being picked and prodded on that end, but if I knew nothing else about Hamza, I knew he would work through whatever issues he was facing.

As for me, I was continuing to be cognizant of the environment I was in. There was one thing that was triggering a deeper emotion within me than I had ever anticipated and that was the proportion of minorities I was seeing. Oftentimes, we would hear the statistics being thrown at us about the large number of black and brown brothers being incarcerated in high numbers. However, when seen with your own eyes it gives you a more heart-wrenching view of reality. I stopped Hamza mid-sentence as he was talking and I gripped his hand tightly and said, “This is not easy.” It caught him off guard I am sure as he didn’t know what the heck I was talking about. But now that I heard myself say it aloud, I might’ve had him thinking I meant I couldn’t handle him and I being in a relationship. That was far from the truth. Our kiss had sealed the deal that I was going to take this journey with him and allow him the time and space he needed to deal with his circumstances. He looked at me for a moment as he tried to get a clear understanding of what I was referring to as I looked around the room.

“I now understand why our sons and daughters, mothers and grandmothers are suffering at the rate that we are.” He was still looking around to make sure he was following my train of thought as I said to him, “Look around this room. This is where all of our brothers, uncles, fathers and grandfathers are. This is breaking my heart. This system has incarcerated our strength. And it disproportionately has the upper hand on demolishing us as a people.” I say all of this as a young man walked past us who worked inside the visiting room and the mere look of him made my eyes well up with tears as he greeted Hamza with the greeting of one Muslim to another. He was not wearing a kufi at the time, so I had no way of knowing whether or not he was practicing Islam. He looked like he was only as young as our own son. After they greeted one another he continued on with his task of making sure the visiting room was clean of debris, trash bins were emptied and recyclables were separated. I immediately said to Hamza, "He is a baby."

Hamza, in a shameful tone responds with, "Yeah, I know Qiana, but there are far more behind that door that don't come out that look like him and are even younger. This is terrible and no place for them to be. This is why I am trying to stay in my children's lives, especially my boys, to keep them from making the same mistakes I have and ending up here. I have done enough time for all of my children combined. I owe it to them to try everything I have in me to keep them out of here because the system is making room for them to come in and stay for a long period of time. This is also why it hurts me so much when I stopped having the ability to see them; they were no longer able to visit me regularly. Now they are getting older and starting to do things that will land them in here. I know if I had the chance to talk to them more and see them regularly, even from here, I can help guide them away from those paths. I know my kids need me and they'll listen to me when I talk to them. The one daughter I was able to see pretty consistently is doing well. Her mom stayed on top of her pretty much like you stayed on our son, so they are going to be okay. But my first two had less of that discipline growing up and they've had to struggle harder to stay on track. I know I can understand what my kids are all facing and that temptation of the streets and the fast lifestyle. I've walked in their shoes. I know if I don't get more time with them one of them is going to end up in here and that is going to kill me! I pray for them 5 times a day, and sometimes more, to just hold on. I owe them my life and so much more."

We broke from our conversation for a little while to grab a bite to eat. I noticed Hamza didn't eat the things he used to. I thought the change was a rare request when I saw him in Susanville, but I see he wanted the same things here at this facility. He stuck to the cheese pizza and no buffalo style chicken which was his favorite. He respects his religion and only wanted things that were halal. Yes, I was taking mental notes of everything. We purchased the items and the photo line was now shorter than it had been so we jumped at the opportunity to get our pictures done before everyone realized how short the line was. I stood in line while he placed the food on the table and soon we were taking two pictures. It was a bit awkward as I was so used to there being three of us in the pictures. However, this time it was just the two of us and he didn't adopt the standard poses for us in a photo together, but there were subtle hints in our poses that indicated we were more than just our son's parents such as the placement of our hands and we both had smiles, not only on our faces but also in our eyes.

After the pictures were taken and we had our time at the microwave where we made small talk as we waited for the food to warm, we returned back to our seats where we again talked about ourselves. We both agreed that we never imagined over the years that we would ever be at this point. He was busy living his life and I was busy living mine so we did not see our paths crossing with this result. We laughed about it every time we thought about it or mentioned it. One of us would sometimes remark, "Man, we have come a long way from our teenage years," or "Never would've guessed…."

After we had talked for a few minutes he asked me, "Can you see yourself being married to me?"

I try not to answer in a manner that is hard to understand, but the easiest way I could think to respond was, "Hamza, I don't want to be married in this place. I hate EVERYTHING about it!"

He looked at me with a disappointed and let down face. I hadn't succeeded in giving him the answer he wanted. I was not trying to say I didn't want to marry him. However, I meant the part about not being married in prison. I tried to clean my response up and could only hope it worked out better if I explained what I meant. "Listen, I want to marry you; that is not my issue. I just don't want to be married in this place. I want to be married by the water. I want to enjoy no restrictions and not have all these people in here that I don't know, staring at me. I just want a different environment, not a different person." It was my effort at being more detailed, but that didn't go over well either.

He looked at me and said frankly, "This is all I can offer for the time being. I have nothing and no money, although I am not the man who still sells drugs, has cell phones, runs scams and all the other things that were causing me all the trouble they come with. What you see is what you get. In the past I tried to wine and dine everyone and buy all of their attention and affection, including my children, and they were all readily available when I was doing all the wrong things. Then the moment I decided to change my life and do things according to what is right this was when everyone started dropping off. So it hurts a bit when your initial response is, 'not in here'. I understand what you are saying and you are right. You deserve the wedding of your dreams, but I am asking you about marriage. That's the part of the relationship where there are no dresses and pictures; it's the commitment, connection and union that a couple makes before God. You are what my right hand possesses."

I was stunned by what I had just heard come from the mouth of a man that I once didn't even take seriously. He was just a 16-year-old teenager that I used to enjoy only in the privacy of my home who had a Caprice Classic with the music blaring down the block. Now, he was a soon-to-be 40-year-old man who had side-tracked my entire train of thought as he reached into a topic beyond the surface and was digging deep into my intellect. I could jump right in and give him a detailed response that would capture him as well, but instead I reserved the right to just take in the truth that he had grown, matured and was willing to offer me far more than just the status quo. He was interested in establishing a foundation built with substance. So I gracefully requested that I answer him more thoroughly at a later time and date. He honored my request.

I took more time to enjoy his presence and blocked out the room full of people, shuffling back and forth. Soon our visit would be coming to an end and we would be required to form a line, receive our passes on the other side of the grill gate and process out. He lets me know that he was expecting a visit from his wife soon and when she comes he would be sitting her down with her at the opposite side of the table from him, to attempt to speak to her maturely about the changes that would be happening in his life, hoping that she would be willing to part ways amicably. However, he didn't expect it to be easy but he remains optimistic. He reassured me that this had nothing to do with me and the breakup had been a long time in coming. He had tried to do all he could to give their marriage his full effort, but there was nothing left. He informed me that over time he would be letting everyone, including his children and their mothers know where he stood with the looming divorce and hopefully, as he gave me the side-eye, his upcoming marriage.

I gave him a grin in return and sighing coyly I replied, "In due time, mister."

We both chuckled as we were interrupted and notified it was the end of visitation and we must follow their instructions. We formed a line and soon we were at the gate where he pulled me into his arms with his hands planted firmly on my butt and my arms draped around the back of his neck, and again he pulled me close against his hard body for a passionate parting of my lips and a slip of his tongue to intertwine with mine before we separated and gave the appropriate greetings of Islam. He said, "I'll call you soon and I love you, Qiana."

I responded, "Okay and I love you, too, Hamza." For the first time it was an altogether different feeling as I was leaving. This time I was parting ways from someone I wanted to share all my time with.

The first visit was memorable and lasting. I would make a number of trips to visit after the first one. It was now customary that we set aside time to see each other as our interest and relationship was in forward motion. I was getting to know Hamza on a level different than I'd ever desired to get to know Vandrick. The difference between being co-parents and being in a romantic relationship took a far deeper personal commitment and dedication. Co-parenting required far less of an intimate connection between us and took more of me just making sure that he and our son were able to remain connected. Now, we were becoming more vulnerable and letting down our guards, revealing layers of ourselves that had never been discovered. At times it was easy to do because we had shared and opened ourselves up to one another to some extent over the years. However, taking advice from one another, that was deeper than either of us had ever been. I respected his candor and appreciated him as a man for allowing me in. And, it did demonstrate that he was serious about what he was putting on the table for us.

But, there was still the situation that he was working through with his wife and just as he had predicted it was not an easy process to go through. She was making every bit of a ruckus that she could think of, but he had managed to not have us bumping into each other in the visiting room. That would not have gone over well at all! I had my reservations and mentioned them from time to time to make sure he was 100 percent certain that I did not want to be placed in any awkward situations. He was confident that was the one thing I did not have to concern myself with.

He made it plain by letting me know, "I have no problem with sitting you down in this visiting room with her, or any of my kid's mothers, on the opposite side of the table from us and letting everyone know at the same time what is taking place. But I also know the only person that will show up on that day will be you. No one wants to sit before you and me and hear the truth. They have heard it from me already and for it to be stated bluntly in front of them would only force them out of their false sense of hope they are trying desperately to hold on to. No one is going to dictate the choices I make in my life. They can either accept it or they can remain upset, but it has no effect on what I am doing with my life. I have done my best in the relationships I had while I was in them. And just as I had to accept the choices and decisions they made over the years in their lives, they will have to respect mine."

I listened attentively to the statements he made and allowed him to release whatever emotions that were pent up from what he had endured to be revealed without rebuttal. One thing was for sure, we all had a past that would define us as individuals and where we wanted to be in life. Those circumstances were a bit awkward and touchy as we maneuvered through them to the best of our ability, but I assured him I would give him the time and space he needed to get through what he had to do, even if it meant he had to try last ditch efforts at salvaging any relationships he felt are worth it. He guaranteed me that he knew exactly where he wanted to be and who he wanted to be with and I should be without a doubt that it was me.

CHAPTER FOURTEEN
CO's Are Not Your Friends

I let him know being in the visiting room came with its own layer of discomfort because there was definitely no privacy. I sarcastically referred to it as the "fish bowl" because everything was on display. I felt like a goldfish any time I was in this place. Even as I processed out of visiting early on occasion due to multiple things scheduled on the same day or our visits were terminated prematurely so other families could visit with their loved ones due to overcrowding, there were the looks from other routine visitors, staring and trying to guess the story behind, who were you? Who were you coming to visit? How long would you remain? Or, Are you a trafficker? The whispers among the small cliques and sly looks were obvious to even the most oblivious person. They were trying to figure you out or think they knew and had already made an assumption they were satisfied with.

Even the Officers had a way of letting you know they could see what was going on—from a slick comment they might slip in, to a questionable gaze with a smirk, or even laughing outright with other Officers as they watched the multitude of visitors and the patterns being established. In the short period of time I'd been visiting, I watched an Officer use her over-familiarity with an inmate go too far as she interrupted his visit to hold an impromptu conversation with him as she rudely thrust her butt into his visitor's face. It was clear this was not a conversation that was in any way urgent or a corrective action necessary for her to approach the inmate. Instead, it was a great opportunity to let his visitor get the subliminal message that she was in control of that inmate, in more ways than one. If that didn't speak volumes to me to be mindful of the environment I was in, I don't know what else would! Some of the Officers were professional and it was reflected in their work ethic, but some were questionable to say the least and definitely needed to be monitored far more closely and reminded of their boundaries in their line of duty.

That scene took a while for me to process and get back to my own visit because it blew me away, but that was only one of the many times that I had been shocked beyond my belief. I had seen officers stand by passively as inmates' Saturday and Sunday visitors would show up on the same day by accident which caused full-blown Jerry Springer moments in the visiting room, drawing every eye toward the mix-ups on the visiting room floor. There would be shaking of heads, mouths and eyes opened so wide you could see their inner thoughts; kids distracted from playing on the small structure that was in the children's play area as they tried to figure out who was being louder than they were. I have determined this was not something I had the patience and tolerance to withstand and if this was in any way the type of embarrassment I might face in my very new relationship, Hamza would never have to worry about me ever making available time to drive up there and walk into this visiting room again. I respect myself and my time way too much to put up with the things I had been a witness to in this fish bowl.

I saw the visitors who tried to become friendly with the Officers and it was entertaining to see how fast those efforts went downhill. I guess those visitors lost focus and forgot that this was the Officers' jobs and how they maintained their livelihood. More often than not they would not jeopardize their profession based upon becoming friends with a visitor or an inmate. However, some would make an attempt to try to gain that leverage with the staff. The prison would try to curtail that type of interaction by routinely moving the staff around so they wouldn't become fixtures in any one location, but sometimes it took a little while for the staff changes to take effect. I was the same as I had been growing up all my life. I wanted to be the least noticeable so I kept conversation to a minimum. I had always been the person that was on a "need to know basis." I was cordial and respectful, but I had barriers built around being trusting and friendly. I never wanted to appear as if I was gaining or had access to something that I didn't. Plus, I understood the rules of perception. So I responded when spoken to and definitely followed a rule established by Hamza from day one of visiting: "Never get too friendly with anyone at this prison, both staff and visitors. You don't know what folks are into. Also, never let anyone get into your car." I respected those cautions because I knew they were in my best interest.

I saw some of the most innocent looking visitors, mainly females, being carted out by an Officer and the inmate they were visiting placed in handcuffs by the "Goon Squad." Those were Officers hired to be a part of the Investigation Unit. They were the ones who would take the visitor behind closed doors to confiscate the evidence. This all happened because the visitor determined it was okay to take a chance at bringing the inmate drugs. That was definitely one of the things they would never have to worry about with me. I had way too much to lose and way too much at stake to jeopardize it all behind drug transport in the hope of maintaining a relationship. The request alone would make me lose all respect for any relationship I was in. It would be a clear sign that he didn't give a damn about me! I was just collateral damage in the bigger scheme of things. My freedom would have been at risk just as his had already been taken away. Yet, I saw it happen multiple times. Sometimes it was a successful transaction and sometimes an epic fail for both parties involved. I gathered that the times the Officer's came up short, the Goon Squad would go back and study where their mistakes were made so they could eventually catch more folks in the act.

There was one unfortunate time when Hamza and I were suspected of making a drug transaction. It had been a stressful time for the "Goon Squad" because they had caught an inmate in the weeks prior. However, the visitor was successful in getting away. So they were watching all things more closely. I assumed after they failed to catch the visitor their egos were a bit bruised because she was able to escape the facility right under their noses. So with Hamza and I, we were visiting and going through our normal routines: he would come into the visiting room after

I had checked in and had gone through being searched and passed metal detection screening. He would greet me with a hug and a kiss then we would sit down and pray before we did anything else. After a bit of small talk we would go from vending machine to vending machine and select items to eat during the visit; even our food selections had become routine because what we chose was pretty consistent. We would sit down and talk as we ate the items that required no warming and later, as the visiting room started to settle down from all the movement, we normally warmed our food in the microwave at that time.

Hamza and I started discussing the events that happened in the weeks prior with the inmate that had been caught but the visitor got away when we noticed we were being watched more intensely than usual. We even mentioned it but didn't think much about it and brushed it off. We knew there was nothing they could ever bring to our attention because neither of us engaged in drugs or any of the other shenanigans that warrants all the extra surveillance. We would laugh at how amateurish their level of training in monitoring was, but I took notice of them looking directly at us and pointed it out to Hamza. The visiting room had settled down and the microwaves were less congested so we decided it was a good time for us to warm up the cheese pizza he was having and he would take a bathroom break. We headed in those respective directions and he told me he would come over to the microwaves where I was right after he came back from the restroom. When he returned we stood there and talked a while and then took our food to the table where we were seated, facing the podium that the officers sat at with their computer monitors to watch all movement on the visiting room floor from the cameras overhead throughout the entire room. After we blessed our food and began to eat.

I looked into Hamza's eyes because I could hear the keys jingling from the Officers movements. I noticed that the Goon Squad was already present in the visiting room, but the look in Hamza's eyes communicated to me that the Officers were walking toward the vicinity of where we were sitting. I never looked up or turned around. I knew I could watch him and get a general impression of what was taking place. Nevertheless, I was shocked when five Officers stopped at our table to my right. Two were Goon Squad members, fully dressed in black and three were regular Correction Officers in their beige and green uniforms. One was a female Officer which was standard procedure since she would need to search me if things went that far.

Without hesitation I hear the command given to Hamza, “Stand up, turn around and place your hands behind your back.”

I think for a temporary moment I forgot my persona of remaining cool because just for a moment I looked up and immediately responded, “What the fuck did you just say?” I knew I heard what was said, but it was my gut reaction. Hamza and I had talked about this many times and on various occasions we saw inmates being escorted out of the visiting area. Handcuffs definitely implied drugs were suspected. I am floored that they believed he and I had anything to do with drugs! So, Hamza followed their order. He placed his pizza back onto his plate, stood up and turned around. His calmness helped me to pull it together because, trust me, the girl raised in East Oakland wanted to act a fool in this place right now, but I must remain focused and not give them what they were looking for. I was told to go with the female officer. So, I stood up. I was not handcuffed or going through anything as extreme as he was going through, but I took the one item I had, which was my clear bag, off the table and walked towards the grill-gate with the female officer walking alongside of me.

We stood at the grill-gate and waited for it to open. I thought to myself, "*Oh they learned this trick from the last incident. Close the gate first so the visitor won't get away.*" The gate opened for us and I was told to take a seat on the sofa that was in the waiting area for visitors who visited inmates behind glass and non-contact visits. Once seated, the female Officer wanted to make small talk as she said, "I was just told to stand here with you, and then I will be informed of what is next."

I looked at her and replied, "I was only concerned about when I would be able to pee. I was holding it in there."

She looked at me and said, "Well, hopefully this won't be too long. I know they have to search him first."

I asked, "So you all think he received drugs?" It was a ridiculous question on my part, but I was still in a state of shock about what the hell could make them possibly think of us as suspects, especially since there were so many cameras in the visiting room. I had assumed they would watch us on camera for a while before being quick to judge, but I guess not.

Then she said, "Well, they thought you put something in the chip bag and he was going to get it out!"

I looked at the Officer and laughed. "Well, you all were definitely practicing because I did put something into the chip bag. It was the napkin I used to clean my hands." I think to myself quite often so my thoughts were, *"Your kindergarten sized tables that we share with another inmate and his visitors is not large enough for us to eat our food plus have a pile of trash scattered all over it, so I consolidate as much of our trash as possible to accommodate others."*

She looked at me as if she read my thoughts and tried to clean up her ridiculous statement with, "Oh well, some visitors and inmates will flatten out all of their trash to make sure it never appears to have anything in it."

I clearly had an "Are you kidding me?" look on my face because I know I wear my expressions on my face. I had been visiting prisons for a while, many years to be exact, and I had never witnessed this extra measure taken. So, I let her slide with this new level of bullshit in order to help her justify this craziness. But I did hit her with a bit of extra information so that she would get a clearer picture of me for her and her staff. "I have way too much to lose to come in here and transport anything because my years in service matter more."

Her eyes widened because she must have assumed I couldn't possibly work for the government, be educated and visit an inmate regularly. She asked, "You're in the military?"

I looked at her with a matter-of-fact grin on my face and replied, "Roger that!"

She was clearly stunned and tripped over her words as she said, "Oh, oh…thank you for your service." She started listening to her radio where a male voice was stating they had come up with nothing. She told me it would only be a couple of minutes more because the investigating Officer wanted to speak with me. Then I'd be free to go to the restroom and return to my visit.

The Goon Squad Officer came into the room from a side door and walked over to the sofa where I was still sitting. The female officer was standing slightly to the left of me. He asked if he could speak to me in the hallway and I stood and walked with him in that direction. I didn't mind because it was closer to the restroom and I still had to go. I stood with my back to the wall.

He said, “I apologize for any inconvenience we caused you and your visit, Mrs. Jones. We just saw you put something into the chip bag and we had to follow up to make sure no drugs were being transported to Mr. Jones. I hope you understand. He is extremely concerned about you and how this mistake may affect your visits. We apologize again and if you have no questions for us, you are free to go back and continue your visit. ”

I acknowledged his apology with a simple, “I have no questions aside from am I free to go to the bathroom?” I was sure the female Officer would update him on anything she and I discussed while he had been in the back with Hamza.

He replied, “Absolutely.”

I excused myself and went to the ladies room. I was no longer in shock, but I was frustrated. I didn’t want to let this ruin our visit as I returned to the visiting room after they opened the grill-gate to let me back in. I saw Hamza at the podium with the Officers, about to make a scene while asking in a loud, agitated voice, “Where is she? Did she come back in here?” I fixed a smile on my face so that he could immediately see that I was okay. The Officers were equally excited to see me walk into the room. They knew my missing was about to be the straw that broke the camel's back for Hamza. He was not concerned about himself, but when it came to me, that was a whole other pony-and-dance show.

The Officers started pointing in my direction. "There she is, there she is," they said, trying to quickly diffuse what was about to become a situation.

He quickly moved in my direction and gave me a hug. As we walked hand-in-hand back to our seats, he said, "I swear I wasn't worried about me. I know these folks were going to do what they wanted to do with me, but I was worried about you and wondering was this going to break you?"

I held his hand and replied, "I know how to handle any form of integration, remember, so don't worry about me. I can handle this. I knew we were not in the wrong, so I am okay. This mess doesn't bother me like that. The only part that threw me off for a quick second was them putting handcuffs on you. When they told me to stand up, I was okay. But, when they told you to turn around and place your hands behind your back…well…I immediately started cussing, but your calmness eased me."

He said, "From the time they headed in our direction I started praying to Allah. I asked for patience and peace to remain with me because in the past I would let things like this get the best of me. I would have ended up in the hole and up to be transferred out for giving these people just what they wanted to see from a person like I used to be." He paused and continued, "These areas of my life are being tested and I am getting better at dealing with them. The only person I had on my mind was you. I didn't even want them talking to you. You don't have anything to do with this place aside from coming to see me."

We comforted each other and finished our meal. We discussed the information the Officer tried to sell me about flattening out potato chip bags and other trash. We both called that bullshit and would make jokes about it from then on out. But we had each other and that mattered more than the humiliation they had attempted to cause from their mistaken jump to judgment that had been shared with every visitor and inmate in the fish bowl.

That incident left a lasting impression on the both of us, but moments like that also forced our relationship to be tested and strengthened. We never knew what caused the Officers to focus on us in the first place. Was it suspicious activity, an investigation lead from another inmate's wrongdoing, or an outside tip from a bitter ex trying to disturb our time together? All the drama we faced only taught us how to communicate more effectively, depend on one another more strongly and pray that much more intensely together for guidance and relief. It had not resulted in our losing sight of the path that we're on, or break under pressure. As we both acknowledged, those trials had built us up and whatever we're facing we're learning to tackle things from a different angle. Anytime we're forced to spend too much time apart, we agreed we must make an extra effort to break that distance, either by more letters and cards, more phone calls, or a rearrangement of my schedule to get a visit in.

The Goon Squad did try one other incompetent attempt to invoke their presence on the visitors and inmates which ultimately caused controversy throughout the prison and drew attention up through the ranks. One day when visiting had come to an end, the sally port gate for entry and exit to the interior prison grounds broke. No visitors or staff could leave. It was a hot day and the facility was located inland so any coastal breeze was not felt at all. It was dry heat and the visitors had no place to go aside from standing outside and waiting until they could find someone already on the inside of the facility who was qualified to fix the stuck gate. Some thought it would be wise to have the visitors go back into the visiting room and be seated until the gate was functioning because there would be shelter from the hot climate and snacks available to consume if needed.

However, on this particular day the Goon Squad decided to do a mass bum-rush of the visiting room with their K-9 dogs and had all the inmates lay down on the floor. They cuffed their hands behind their backs as they searched for contraband of any sort. This was the first time I had ever witnessed anything this crazy in my many years of visiting any prison facility. This was complete insanity and it was the most utterly ridiculous scene ever on both sides of what was going on. I am grateful none of my children ever witnessed anything like this. I felt bad for the children that were present who couldn't be protected from those antics; and to the disabled and elderly visitors who had to withstand the conditions until there was a resolution. From where I was standing I was able to lean against a guardrail as I waited. In that moment I was thankful for the military that taught me patience from the hurry-and-wait mantra we had become accustomed to. My training constantly kept me prepared that circumstances could change in an instance and it taught me the ability to stand for extended periods of time without it breaking me mentally as well as to withstand extreme climates to the best of my ability.

So, I calmed my nerves as I found a position that was suitable for me. I looked to my right and I could see Hamza's familiar big frame near the window. I could see him lying down on the floor next to the vending machine. I was relieved to know exactly where he was in the building. At this moment I had to be selfish because he was my only concern. I could see Officers moving around and about the inmates on the floor. I could even hear their voices inside, echoing out to the exterior of the building where I and the rest of the visitors were located. The impatience of the visitors started taking a toll on them after a while because some were concerned about their loved ones laid out on the floor. For some it was too far for them to see into the building to know where their loved one was in the mess. Some were concerned about the gate that was not unlocking and they seemed to have no staff responding to help correct the issue. Almost everyone was concerned about the heat. I could hear the conversations surrounding me of people planning to contact the warden as soon as they were able to get out of this situation. Others were deciding they would be notifying the Governor to make sure issues like this were addressed on a more urgent scale. And some had no idea what to do other than to complain in the moment and hoped it would all be over soon.

We spent approximately one hour locked in and the only resolution that one Officer gave about half way through the ordeal—a visiting sergeant—was that there was an alternate way out of the facility, but only one or two people could leave at a time. The problem: we would have to walk back through the visiting room to get to the alternate exit; but the visiting room was occupied by a major shake-down of the inmates in full effect. So we were stuck until finally, the vendor who supplied the food in the vending machines became the hero of the day for us. He removed the face of the gate's locking mechanism, released the latch that had us stranded and held the gate open. The visiting sergeant's staff allowed us to exit through the gate and process out of their facility at the standard rate of five at a time.

I was relieved to be able to make it to my car where I spent a few moments reflecting on the fact that I would never want to be confined in prison for any extended period of time. I knew I had been in the military and it had so many similarities in experiences to being incarcerated, but at least in the military I could look forward to the breaks, leave requests, scheduled vacation days off to give me that time to breathe, regroup, get away and return renewed and refreshed. That was clearly not a possibility when serving a prison sentence. I took a deep breath and thanked God for keeping me from going down the road to prison in my life. I realized that anyone could be just one bad decision away from occupying a cell at any given time in their lives depending upon the situation. Then every aspect of your life will be under the control of a system ruled by Correctional Officers who are paid to provide a cruel service that some would uphold and some would disregard.

CHAPTER FIFTEEN

Going On the Inside

Life had a way of moving you into positions and directions that were unexpected and unplanned but necessary to understand your growth and full potential. I have graduated from college and during my time in the military I had learned about another college I could attend in Oklahoma City that would set me up nicely in the direction I would like to go. This would be another big leap and sacrifice toward pursuing my ultimate goals in the medical field. I was trying to see how I could make it all work with my son already off in college and my daughter in middle school. It was a great opportunity for me, but I had to consider their needs as well. The timing wasn't perfect, but in my life experiences there had never been such a thing as perfect timing. I was always shuffling ideas around some major pros and cons and now I had a new relationship added into the equation. The one thing I knew for sure was that along with all my decisions I had made in the past and my consistent prayers, I would for the first time in a long time take into consideration Hamza's opinion as well.

The course was not a long one, but it would require me to move out of state, possibly moving my daughter as well. She was not too fond about this as her friends mean a lot to her at her age. I remember feeling the same way when I moved away from my friends around the same age when my mom decided to move away from our old neighborhood. I thought this was devastating at the time, but I eventually formed new friendships and slowly lost contact with the friends I had established bonds with and gone to school with. As I discussed the ramifications of this move with my daughter I reminded her that technology was a great asset that I didn't have the luxury of having when I moved away from my friends. I told her she and her friends could stay in contact far better than I could at her age because we only had a landline phone or mail and even three-way calling was a new thing during that time. None of those methods helped me remain in contact with my friends for very long. At least through social media she could see everything her friends were doing and they could see her, too, since they post everything from breakfast to bedtime, especially on SnapChat. She was building up the desire to take the journey with me. One thing was for certain, she still didn't like spending long periods of time away from me after the two years we were separated and the few times I was still required to pack up and go out on orders.

Then plans changed. After being in college for a year my son decided he didn't want to return for a second year. The distance was too much. And, he had also gotten into a major accident which rolled his vehicle, during an impromptu trip to hang out with friends at a college in Las Vegas. Now, his brand-new Jeep that he got as a gift in high school was totaled and he was without a car for a short period of time. The least of my concerns was about the car. I was just grateful after receiving the information about his solo car accident and the pictures that followed that he was alive! He had only suffered a small cut on his hand from the shattered glass. The vehicle was fully insured so it would eventually take care of itself. However, my son was headed back home to California to be closer to his family. I didn't argue or disagree with his decision; even though I wished he would push himself through to complete his degree, I didn't make a fuss. He was getting older and it was up to him to make these types of decisions. And, I was not willing to invest in an education that he was unwilling to pursue.

I had transitioned completely over to medicine which was the initial plan from the time I returned home from the military. That had finally taken place. I was no longer in the mindset where I was pursuing one profession in administration and everything else focused on medicine. Things were finally coming together and making more sense. However, I needed to delve deeper into medicine and the place where I was working I could only work part-time based upon my prior school schedule. I wanted to give myself various opportunities and follow along the path that was drawing the most interest for me. It was time to make a decision—do I go out of state and attend school full time, or do I stay here in California and continue to work part-time plus attend school which was taking longer? California was definitely getting away from bridge programs and leaning heavily towards longer timeframes in school with the goal of racking up more money in the end. Many of the requirements for didactic and clinical studies would also affect my current job; the hours I was scheduled to work were now the hours I would need to dedicate to school.

I was definitely in a full transition period and all the cards on the table would take some sorting out to accommodate everyone that was involved. Plus, life has continuously proven for me: "When it rains it pours." So, with everything I had left to juggle and keep going, I submitted an application to the State of California Corrections Department to work and get my degree as a nurse. I had never thought about it in the past, but I knew a coworker who had worked for them and he provided me with the work requirements and the hours and shifts that were available. It seemed doable because I needed more available hours to work and he even let me know—because his wife was working at one of the facilities—they were providing growth opportunities for their nursing staff to pursue their education by piloting a program called 20/20. In this program nurses can work 20 hours, attend college full-time towards their degree and would still get paid for 40 hours of work. This was perfect for a person such as me who needed to maintain full-time employment while I finished my degree.

This option was the best of all because it gave me everything I needed and I could keep my daughter in California which would be one less battle with her leaving her friends as well as her dad who wanted her to remain close to him. I might even be able to accommodate Hamza in this as well. I found out there was a facility with an opening in the city he was in. I submitted applications to three prison facilities and got replies from two. One was from Tracy, but it was intermittent work and the other was from Vacaville with full-time hours, although it was a limited-term contract. Vacaville was the best choice of the two and they seemed really eager to hire because the interview was scheduled rather quickly. I was interested and took the chance to make this happen. However, I had only one reservation about their hiring process as I thought to myself, "*Will they hire me even though I have a child by, am currently with and routinely visit an inmate at one of their facilities?*"

I attended the interview and even though I was a nervous wreck during the process, I did well. They had sets of specific questions to make questioning an even playing field for every applicant who was interviewed. I left feeling confident that I answered each question thoroughly and to the best of my ability. I only felt unsure about two of my answers, but that was still rather good. I would wait patiently to see where things would go from there. I was leaning heavily on this being the path I would take if this all came through. My daughter and I would have to move a little farther away because the commute would be too far from where we live now, but it was still better than moving to Oklahoma. She was not too far from her dad and her friends if she wanted to visit either of them. She still felt “some kind of way” about leaving her friends, but it was less traumatic than the original plan.

The interview was a success! Within a few weeks I had gotten a tentative offer to start the process of becoming a full-time nurse for the State of California Corrections Department. But, before I agreed to anything, I again asked the one question that had been looming over everything, even though it was addressed on my application and additional paperwork. I wanted to be certain there would be no conflict of interest as I stated to the Human Resources Administrator when she offered me the position. I asked her, “Are you aware that I know and have a relationship with an inmate at the neighboring facility, California State Prison Solano? Would this become a conflict of interest if I took this offer?”

She asked, “Did you disclose the information on your application?”

I replied quickly, “Yes ma’am, I did.”

She answered, “That is all that is required. You have been honest in your application process and you identified who he is. We in no way can discriminate against you in the hiring process based on your relationship with an incarcerated inmate. You are completing the hiring process for a facility that he is not housed in so there is no conflict; if you were to apply at a facility where he is housed, that still does not deny you the right of employment. You are the employee and our process stipulates we will move him to a different facility so that you both do not have direct contact. However, I do thank you for your honesty and getting clarification because some people will hide that information from us in the belief that first, we will not find out and secondly, they often assume we will not hire them based on their knowing an inmate. Again, that falls under our rules of discrimination. We cannot do that."

"Thank you and yes, I will accept the offer." I told her with a smile on my face.

We made arrangements for me to continue with the remainder of the paperwork to complete my hiring requirements and I was finally able to share the great news with Hamza. This was a big relief for both of us and was the final determining factor in making the decision that would be best suited for the needs of my family and all involved. I even started the process of applying for a college in the immediate area of the facility as well. Things were looking optimistic.

The good news that I was hired was what I needed to keep us motivated on this journey. Hamza was still going through hell in dealing with his wife who was unwilling to quit without every stone being thrown. He had tried talking to her rationally; giving her the divorce papers in the visiting room and even mailing them to her address where after he called to verify she had received them, she informed him, "I threw them in the trash. I am not giving you a divorce!" and hung up the phone in his face. Hamza was trying everything he could to handle it all without my involvement, but that wasn't possible for long. He and I went through a period of time where we separated so he could just focus solely on his issues with her and I could focus on the things that I was going through and the stress I was under. This was a period of time when my tolerance for all the nonsense was far too much for me to bear. Our son was going through his growing pains and I was mentally drained trying to make sense of everything that was falling into my lap. I needed some time to myself.

Hamza kept track of the days that we were apart and after our necessary separation we were without any contact or visits for a total of 66 days. But as we found out, when it was time for us to come back together we both reached out at the same time, writing each other letters that would unknowingly cross in the mail as he spoke words from his heart to me and I shared words from my heart with him. We both agreed it was time to get back to communicating and spending time together again. As we returned back to visiting, he shared with me what the time we spent apart and the distance did for him. He told me he had to tap into a deeper level of strength as he thought he was likely to be alone for the remainder of his sentence because he still was not making any headway in dissolving a marriage that was unhealthy. He vowed he would not continue down that path just to fill a void, or stay married for the title and the relationship lacked everything else a marriage included. I shared with him the pressure our distance also placed on me; the things that I needed time to reflect on and sort out; the growth we needed to work harder at communicating and listening attentively to one another. I let him know we were not in this relationship to determine who was more right or more perfect. We needed to practice working towards resolutions. This was new for both of us because our prior relationships dictated certain requirements from us and we both needed to actively work to get rid of those bad habits.

Our first visit after the separation was refreshing and it was definitely a time for us to talk through many things and open our eyes to areas that needed addressing. In order to move forward we had to get through a lot of our past. We realized we deserved this chance at what some labeled as real love. After a few visits, Hamza shared with me that he needed my help. His wife was not going to let him go even though he had presented her the divorce papers during a visit as well as mailed them to her at her home where she threw them in the trash. He had asked her over and over again for a divorce and at this point he no longer made any attempts to contact her. He just wanted the divorce papers served to her again so that his divorce could move in a forward motion. He knew I could help him in this area because I had prior knowledge of what it took to get a divorce in California.

So one morning during visiting, he showed up with a manila envelope in his hand. After we sat down, he showed me his divorce papers and he went through it page by page and line by line, covering every area that needed to be signed. Once he was certain that everything was recorded in the papers properly, he let me know that he would place them in the mail for me to receive. He needed me to take them down to the courts to be filed and soon enough she would be served by a Process Server. I assured him I could complete that task for him. This was also the same day that he decided, without a doubt, he no longer wanted her to have access to him and he would make sure the following business day to have her removed from his visiting list. There would be no woman that was going to interfere with the journey that he and I were on. He appreciated my patience, but he knew I deserved more. And I knew I deserved his full attention and commitment without any further distractions.

Soon afterwards, I no longer worked for my former employer and was scheduled to start orientation at the neighboring prison, California Medical Facility. I moved to the city and my daughter started a new school there. We were both getting acclimated quickly to all the changes that had taken place. We had our share of feelings about the transition we were going through. I loved the work environment that I had left. I worked with a great staff and I loved the patient population that we served. So working with inmates would definitely be different. My daughter was going through her own emotions of missing her friends as well and she made sure to let me know at every available opportunity. After a while, those sentiments had started to be made less until her previous school would do something she considered cool that her new school didn't do. I tried to accommodate her to the best of my ability by allowing her to occasionally go back to sporting events and dances as a guest and to hang out with her friends. I didn't want to make the move unbearable for her. But where I could provide some leverage I did with the help of her dad and my grandmother who still lived close to her school and friends.

Working in the prison was definitely different because a major issue addressed repeatedly was "over-familiarity." We were to keep a safe and professional distance with regard to communicating and interacting with the inmate population. We were taught about things to say and do to not give an inmate too much information that would later be used to involve you in a swindle or where they could apply manipulative tactics. We were to make sure we never compromised our employment by bringing in anything that would constitute as contraband, especially cell phones and drugs which were high commodity items. I never shared with my co-workers my relationship status or that I was familiar with the prison from a different perspective as a regular visitor. I had to adapt myself to the environment of being on this side of the concrete. I would determine who to be friendly with as time went on. I still kept my mindset that information was on a need-to-know basis. It was already disclosed in my file, so the entities that needed to know were fully aware.

I knew Hamza had no desire to ever be moved over to this facility, so I had no concern that he would come here. This facility mainly housed inmates with serious medical conditions and gratefully he didn't fit the criteria. But, he was also under the impression that it didn't have a "main line" like a general population facility had. So coming to my facility was not a choice he'd willingly accept. Plus, Hamza had been on a strict path of turning his life around and doing what he needed to do towards getting his life back on the right track; staying consistent at it so he could be prepared for the day when the gates open and it was his turn to go home. He had been committed to this process for about five years now. He had been accepted into a first-generation honors program established at Solano called DeLancey Street. He was transferred there on the same day I was hired to work for my previous employer and his transition worked out smoothly for him. He had access to multiple rehabilitation programs, counseling, vocational skills certification and on-the-job-training. Moving away from all these opportunities was not something he was willing to lose access to.

California prisons were going through a structural transition, focusing more on reform and lowering the previous high numbers for recidivism; not to mention they are faced with a bigger problem of overcrowding and now were responsible for paying the Federal Government fines until they could get their numbers down to tolerable rates. Due to mass incarceration, inmates were literally sleeping on top of one another, three bunks high in gyms and anywhere else they could find to house them. It was unsafe and unsanitary conditions for both staff and inmates as they tried to control and contain those numbers. The Federal Government had no choice but to get involved and force some corrective action, especially as they looked closely at the number of people being given long sentences that didn't fit the crime of non-violent offenses. Something had to be done.

Once I had completed orientation I was ready to start training in the units. I shadowed employees to gain firsthand experience at working in a prison. I was initially uneasy because I was more concerned about working so closely with the inmates and all the things that could go wrong after the stories we were told. I was not concerned about being "over-familiar." The last thing I was looking for on this job or any job was a man. I had one of those and as a matter-of-fact he wore the same attire as the men I'm around so needless to say, I don't need another one. Thanks, but no thanks. The Officers had proven not to be my cup of tea as well, so no on-the-job affairs were on my checklist of things to do. I planned to come in, do my work and pick up my check. That would be the extent of my presence there.

I was shocked to find everything in this place was still manual! The medical industry had been automated for too many years for California State facilities to still be processing medical records with pens and pieces of paper. Now, I understood completely why so many errors and potential for malpractice suits were taking place. Human error was at an all-time high at manually maintained facilities. This was my biggest concern and it came with an increased possibility of placing my license in jeopardy which was a risk I was uncomfortable with. I was advised that the facility was upgrading to automation within a few weeks, but it seemed much of the staff was in opposition to this taking place. I was one of the one's looking forward to it. In medicine, checks and balances were completely necessary. I assumed most of the employees had been conditioned to operate that way so change would not be easy for them. It made sense, but safety of the patient was of a higher priority to me.

Soon I would be required to work alone and I wanted to make sure that I was confident in my performance. A supervisor had already delivered a message to us as part of her welcoming pitch. She had said, "I know you are concerned about working in a prison and about the prisoner population. But you already know that they are here for a crime that they committed and you are coming into their home. It will not be *them* you will have to worry about as much as it will be the people you work with to your right and to your left!" I listened and her statement had a lasting effect on me. I would leave for work extremely early and get off mid-day. Before I walked into the facility I spent time in the parking lot, praying. I asked God to cover and protect me from things seen and unseen; to help me stay in my right mind and return home the same; to help me not to grow a callus around my heart which I knew had caused others to administer medical treatment at an inferior level to patients who deserved better; and I reminded myself not to forget to place my cellphone in a preselected spot so I didn't forget it or walked in by accident with it on me. Routines would soon make it easy to remember.

We had been warned about the many hours of overtime we would be required to complete after we made it through on-the-job training. I was not looking forward to that at all because my daughter was here without anyone other than me. But her dad offered that on the days that I was mandated to stay 16 hours, I could call him and he would drive there and stay with her until I could get home. This would be the plan for the time being unless we came up with other arrangements. Being gone from the dark of morning until the dark of night was not ideal when you had a teenage daughter to think about.

I was getting adjusted to the many different places that I might be sent to work. We worked right on the units with the inmates. We did have times when we were locked in rooms where they had no access as we prepared ourselves for later interactions. But I am assured, for the most part, the inmates posed no problem. There were a few that could get a bit unruly, but officers were present and some would make it a point to keep them controlled. I also took my own set of precautions to make sure I was safe. I was always in my military state of situational awareness. It helped in this environment. I even tried to share tips with a couple of friends that I had made after I noticed some things were not as safe as they could be. Like the process that really made absolutely no sense to me. Why were nurses with narcotics walking unattended down the hallways as inmates are also freely roaming the halls? The first time I was subjected to this I felt like a walking target with this clear bag which was far from discrete and clearly suggested to anyone looking for a really good high, "Come and get it! Here it is!" I was never placing myself in that position again without proper security. And if they couldn't provide security, then they would be delivering it to me.

The other part of the job which took me and the other nurses by surprise was our introduction to the Mental Health Department where inmates were placed on suicide watch. Those inmates received a 1:1 ratio of inmate-to-staff supervision. The staff person would remain seated in front of the cell's double-paned, wired, glass windows while avoiding any distractions. They would continuously watch the inmate in the event he attempted to commit suicide. This was a customary practice in the mental health field because at any moment of not properly watching the inmate, it increased the chance of an inmate's opportunity to attempt suicide. In this environment the patient was rid of any potential hazardous items such as clothing, shoes, shoe laces, sharp objects and basically anything they could hurt themselves with.

The problem was not monitoring for the potential of harm as it was the increased rates of female staff being subjected to inappropriate verbal and sexual acts we witnessed over an eight-hour shift. It was known that some inmates would imitate being suicidal to come to this department because they knew there was a large female staff population. They knew there was a good likelihood of getting to sit with a female all day where they could masturbate with no repercussions for the act. We were told in training and orientation that we'd prefer the inmate doing that instead of committing suicide on our shift. After a while the shock wore off and it was so prevalent that you would come to expect it. So while it was happening, you did as much as you could to look above their waistline until it was over and in no time they would either fall asleep or pace their cell, trying to decide whether to masturbate again or not. Many of us cannot believe that this was tolerated in a prison setting and there were no consequences, but it was the case at this facility. I had heard that at other facilities, if an inmate even hinted they were going to make an inappropriate sexual advance toward a staff member, they were charged with the act and sent to the hole. But that was not the policy here.

The introduction to this facility was vague and I learned much of it on a day-to-day basis, getting better equipped to handle the environment myself. I found my previous occupation and growing up in the old neighborhood helpful to deal with the atmosphere inside. It was all too familiar. I had been fortunate to not run into or see anyone that I knew from growing up. That was kind of shocking because many of them definitely chose the lifestyle that would land them in a place like this. However, it did provide a sense of relief that I hadn't been placed in a situation of working and running into a familiar face. I quickly found out that there were some famous criminals that I worked with daily; ones that made the news for months at a time; ones that had turned their crimes into documented stories; and some I would hear their name and it might ring a bell, but I might not know the ins and outs of their case. And I would always tell any staff that had all the details, "I prefer not to know." They would look at me a bit weird because it was common for them to sit around, socializing during down times as they discussed a prisoner or the acts they had engaged in. But, I did not want to know.

I preferred to work without bias. I did not want my judgment regarding their level of treatment tainted by the crimes they had committed, so I went out of my way to not know the rap sheet attached to the inmate that was in front of me. I relied on the skills and tactics that I had learned in the event I was ever placed in a position to defend myself; I will do that. I knew where everything I needed to protect myself was in any room, cell, or office if needed and I was not afraid to use them because my goal was to go home every day.

I adjusted to the environment very quickly. I had not been placed in a situation where I had to perform anything extreme and I was finding out that, for the most part, most inmates just wanted to do their time and be left alone. I was willing to give them just that—medical care and send them on their way. Some were funny and would cause a laugh or two and some were even talkative. Then there were the ones that wanted sympathy and would break down into tears about every little thing. However, those were some of the same things any nurse would run into no matter where they worked. I was finding out this was just another location with a large number of patients, nothing more and nothing less.

I became acquainted with four of my co-workers and shared with them who I was and a little bit about my background story. Well, one of them already knew me from the outside. He and I used to be in the same military unit together. He was actually one of the first soldiers I met and he sent me off to basic training with the encouragement to succeed. He was just returning from the same course I was heading off to, and he told me, "Don't come back without that certificate. It is what you are going for." I would remember those words and carry them with me along with some material he gave me to study from as well. So it was refreshing to see that he was working here and had been here for a pretty long time. He was doing great for himself. All four of them knew that I was engaged to be married to an inmate and where he was located. They shared with me their lives as well. The ones I have trusted with this information had proven that they were okay with the information and were not judgmental at all. It helped to have a few people around that I could trust and it allowed us to be able to communicate without my guard being up all the time. The only question each of them asked without hesitation when they found out I was engaged to an inmate was, “Does the prison know?” That was the most popular question and I was confident when I answered each one, “Yes, it was the only way I was willing to accept the job offer.” Once they knew that they tended to say, “Well then, you have nothing to worry about.”

My schedule still allowed for me to be able to visit Hamza because I had the weekends off through my orientation and training phases. I saw him as much as I could because I knew this was soon about to come to a swift end along with the schedule that didn’t include all the mandatory overtime. The salary here was equivalent to that of any other nursing job, but the reason nurses make so much money in the prison is because of all the hours of available overtime. That was great for a person that was juiced about making a ton of money and liked buying big-ticket items, but for me, my family time and taking care of myself was far more important than that, especially with a teenager at home, often by herself, this overtime did not fit into my circumstances. Also, not having access to my cell phone where my daughter could call me in case of an emergency, I was not comfortable relying on the main office to get the calls to me as she had tried a couple of times and they told her I was gone for the day when in fact, I was still in the Mental Health Department waiting on my relief to show up. This was a bit uneasy, but I would do my best to work around glitches like that because this contract was for one year. Of course it could be extended, but as of now it was a one-year obligation that I had committed myself and my family to.

Hamza shared, during our quality time, that he was happy I did not leave the state. However, at times when he heard how things were run next door he was completely in disbelief because he knew from the number of years he had been incarcerated that what I was saying was nowhere near normal. He was blown away by the things I shared and at times questioned whether or not this had been the best decision for us to make. He knew I would stick it out and he would be my encouragement along the way, but we were both skeptical that this was a good fit. The lack of organization and the safety hazards that existed placed too many things at risk. He knew from an inmate's perspective what was permitted and what was not, so me talking to him from an employee's perspective caused him to always remind me, "Never let your guard down, do what you know is right no matter what, and safeguard yourself from everything and everyone." I confirmed for him this was the same practice and mindset I carried with me every day.

Out of respect for our relationship I had started wearing a hijab during visits and even during my free time as well. This practice became more prevalent during the month of Ramadan a year or so prior and then I continued with it the following year. I am still a Christian and have not converted to Islam because it is not a requirement for us to get married, but I didn't mind being more discreet in the presence of others by wearing it. So now, it is a practice that I have committed myself to. He appreciated it, but said he would also be respectful of my decision if I chose not to wear it.

However, with this added layer of clothing in the prison as a visitor, I was required to go into a separate area and remove the scarf and reveal that I have nothing underneath it that was considered contraband. A female staffer would go inside the room with me and check the scarf for any extra items. Many of the female staff had found that this was a ridiculous extra step in processing, but we all followed the rule because it was protocol. The one staff person who made a comment to me during the search made it as plain as could be when she said, "I don't know why this is necessary since you go through the metal detector like everyone else. The metal detector does not go off when you walk through; just like a woman who wears a bra with metal underwire walks through the metal detector and if it goes off, we do not ask her to step into a room and show us what is underneath her shirt. So I don't get why we do this to you and others. If you pass through the metal detectors and not set them off, you should be fine. This was unnecessary." I had to agree with her insight, but as I knew, the very next time I came to visit I would be required to endure the separate search. It was common as well that the other visitors could see me putting myself back together in the ladies room before I went in to visit, making sure my scarf and under-scarf were secured properly.

They would ask me, "Do you have to be checked every time you come?"

My simple response was always, "Yes." I never knew why the visitors asked and I didn't assume the reason so I just answered the question as simply as I could.

Visiting Hamza had come to an end. It had been a while since I had been able to spend time with him. During those times when I was working, we made it a point to try to schedule our phone calls around a break or if I got off on time, then I would be in my car by the time he called. It was becoming pretty regular for me to work a double-shift on the weekends because that was a popular time for people not to come in for work. I was getting used to it so I pretty much made no other plans for when my shift was over. I encouraged my daughter to make plans to see her dad or visit with her grandmother so she was not stuck in the house all day. I was back to writing more and the phone calls were now very necessary because of the lack of time I had available to spend visiting now that I was off on Mondays and Tuesdays.

As for Hamza, he was hoping his family and his children would get more on board as he tried to re-establish those bonds with them. But, that wasn't working out either so he did things like study Arabic, read the Qur'an, work on college courses, exercise and prepare for his work week to occupy his time. I felt bad because I missed him terribly, but we were both making this sacrifice. There had been an impromptu moment where I was able to make it around to visit him before they stopped allowing visitors to process in. I had to take the chance so I raced around the corner and made it in by the skin of my teeth. I had made it, but he wasn't expecting me. I went through processing and my normal search. One of the female officers who knew I worked around the corner asked me about the lack of security over there. I wasn't expecting the question then she informed me she felt uncomfortable there when she had to transport an inmate over for transfer. I was in full agreement with her and she said, "Over here is nothing like over there."

I replied, "Nothing at all."

Then she said, "They don't even watch their blind spots. Anything could happen. I could never work there."

I reassured her, "I feel the same way. Safety is not their top concern over there. Often the staff refers to the place as Disneyland. They say you guys have 'real' criminals over here as if they don't over there. I just take care of myself."

We both agreed that safety was first as we walked out of the room. I was inspected and headed down to my visit. Hamza was completely shocked when he saw me, but tried to convince me he had a feeling I was coming to see him. He didn't know what it was, but he just knew I would come and see him today. He gave me a huge bear hug and an "I miss you so much" kiss then we began our normal routines of visiting. The visit broke up some of the monotony of us not seeing each other over an extended period of time. Afterwards, we would go right back to not seeing each other again, so the visit was bittersweet. However, we were enjoying what we could get.

We did get the opportunity to discuss the fact that his former wife's games had come to an end. As far as we were concerned we had tried every available option to get her served and she ducked process server after process server and even family and friends who tried to give her the dissolution of divorce packet of papers. However, you cannot hold anyone hostage in a marriage in California. So, we gathered all the information that showed all the failed attempts at trying to have her served and provided the documentation which proved her evasion tactics to keep from being processed, to the courts. With that ammunition in hand the courts granted Hamza's request. After all of that drama his divorce was finally granted! Soon we would be able to move forward with our plans to be married. We didn't want to get married during this hellish year we were going through, so we decided and agreed that the following year would be perfect.

I returned back to my ordinary routines of work, work, work and parenting. My school plans had to take a back seat because I quickly found out that the 20/20 program did exist, but it was not highly encouraged at this facility. Also, it was not offered to limited-term contract employees. Only after my probation was complete and I was a permanent status employee could I be considered for eligibility. If it wasn't one thing it was another when it came to this school thing. I had met every requirement and I was still having a hard time making this part of my life happen. I did my best not to become discouraged as I reminded myself that all things happened for a reason. If it was meant to be it would be.

My plate was always filled so this was no different than any other time. It was not like I was bored or lacked things to deal with because I had also taken on the task of helping Hamza work on his case. This was a challenge within itself because after many years of being let down over and over again, he had reached a point of not wanting to open up Pandora's Box. Going back and reliving all those years regarding events that had happened in the past, I could understand. I was working with him through the emotional battles which came with people promising they would help just to leave him to fend for himself. I definitely knew what that type of disappointment felt like, how discouraging and debilitating it could be. He had made some efforts himself, but at that time he was trying to attain it through illegal measures. He finally came to the realization through his religious beliefs that God would not allow him out of those prison gates through illegal schemes. He was pleased that I was doing this, not to gain his approval, but strictly of my own free will and desire to see him set free. It was not something he had even requested of me, but out of my own passion towards righting the wrongs of the justice system. Also, he was a real friend and I would never want to see a person sit in prison that long for a crime they did not commit.

I had always had a passion for the judicial system all the way back to when I was a young girl. I used to tell my mother I was going to grow up and become a defense attorney. Life didn't send me down that path, but I knew a few things that gave me the confidence and a knowledge base to challenge a weak argument. He knew I was serious about helping him and I was making strides and efforts to pay for and copy his entire trial transcripts, hearing briefs and the marked evidence. Anybody that was sincerely willing to help him could have at least done that much. However, some of the same people that were walking around with chips on their shoulders about my presence in his life were the same folks when asked what they were willing to contribute towards his case? They immediately became silent as they strolled on with their day, living their best life on Instagram and Facebook. It took a lot of research because over the years information got archived and moved around; attorneys did not call back and some had even passed away, but I was consistent in my efforts to help in any way I could. After hearing the details of his case and the lack of evidence, I knew we would soon find the right person who was equally as passionate to fight with us for his freedom. God knew our prayers.

I am excited about the upcoming wedding and I've started preparing myself. In my mind I knew how I would like things to turn out and the family members that I would like to participate, but because this was a prison I would need to minimize my plans so they would be suitable for the environment. I worked with an employee I had befriended and told her the challenges I had to overcome for my wedding. From experience, she let me know that what I wanted was possible. She was a positive and outgoing person with a diva attitude that made me laugh with her a lot. She could be over the top from time to time, but that was her natural personality. She reminded me of family and friends I had grown up around and I treated her as such. I respected her being who she was and her candid banter. We worked side-by-side on the same unit and shared the same responsibilities. She had been with the state for a lot of years so the bits of information she gave me was helpful and greatly appreciated.

The other co-workers I worked with on my unit were far more meek and reserved than she was, but she had a way of even bringing out some of their inner wildness after she spent some time talking to them. The inmates lit up when she was around because she was like one of the home-girls from the block to them. She had a way with words that got them to smile through their pain on those days when they were feeling depressed. She was even able to change their minds about receiving treatments, medications and showers when some of them just wanted to lie around and wither away. Of course, there were boundaries that we all were mindful not to cross as obedience was still the rule, but all of us could appreciate the change of atmosphere that her presence brought. We all knew that when she was around there would never be a dull moment.

CHAPTER SIXTEEN

Work-Life Balance and All the Hell In Between

Another new year was right around the corner and Hamza and I still didn't get to see each other as often as we would have liked, but we were still doing everything that we had promised one another. The days were long and we missed that face-to-face time. Occasionally, he had been able to convince his sister to come and see him. Our son made a few trips to visit as well. It gave them some time to have those father-to-son talks that I didn't need to be involved in. The thoughtfulness of their visits was greatly appreciated because I felt bad that I could not make it. He never made me feel guilty. It was my own thoughts that overwhelmed me. When I told him that in a phone call he reminded me, it could be worse. He told me, "You could be out of state and we definitely wouldn't be seeing each other. We are going to make the best of the circumstances we are under." I had to admit this was true. I am literally minutes away from him instead of a flight or two.

Although I don't get to see him on the weekends, I am always involved in something productive. So we rationalized and made the best of the options that were on the table. During the holidays, my family always called and asked if I would be able to make it to dinner. I guess they hadn't realized this job was much like the military. The number of doubles I was forced to work and being the person with the least amount of seniority, you might as well forget it during the holidays. I was not getting out of this prison without so much as a "low crawl" technique. I signed in early in the morning, before the sun rose, and I would sign out at night, after sun set. If you wanted me to enjoy your freshly prepared holiday feast, put it in some Tupperware containers and I might stop by to pick it up and eat it if I was not too tired. I had to be back up for work in less than five hours to do it all over again. I tried not to complain in front of some staff because they were commuting an hour in and an hour back home so they were getting less than three hours of restorative sleep after pulling a double. I often told the ones I had a friendship with to call or text me and let me know they made it safely home because driving that tired was extremely dangerous.

One friend I would actually stay on the phone with as she traveled if her husband was not awake or at work and couldn't talk to her while she drove. I would hate for her to get into a car accident trying to make a living for her family. In return she had been there for me when I received a disturbing text from my son one morning as I prepared for work. He sounded as if life was getting the best of him. I knew I still needed to make it to work, but I also needed to make sure my son was okay. I couldn't do both so I texted her and asked that once she arrived at work to let the supervisors know I was outside in the parking lot on the phone and I would be in as soon as I could get the situation under control. She was such an amazing person that without hesitation she began to pray for myself and my son. It's rare to find people that knew exactly what to do in the midst of a true storm. Some would give the clichéd response of "I am praying for you" and move on with their day. However, it was heaven sent when, without hesitation, a person stopped all activity and surrounded the moment with actual prayer. When I saw her face-to-face I had to give her my dearest thanks and tell her she meant the world to me. My son was okay for the time being and I made it into work about an hour after I was scheduled to sign in. When it came to my family, my tardiness meant nothing! They would always come first.

I ran into another challenge that I kept to myself and from many people that I worked with. There was a morning that I had to call in and miss the entire day. I had worked a double the night before. I had only been home about four hours and it was almost time for my alarm to go off. I hadn't even made it to my bed. I had fallen asleep on the couch with the television on. My phone rang and it caught me off guard. I woke up and in my half-conscious state of mind I grabbed the phone, but I hit the wrong button and the screen closed. I got a quick glimpse of my son's name on the screen. I am fully conscious and alarmed that he was calling me in the wee hours of the morning. I instantly dialed his number back, but I kept getting his voicemail. I hung up and was about to call again when his call came through on my end, this time I answered in time.

"Mom, momma, mom!!" he shouted.

"Yes, what's going on?" I replied with urgency.

He cried out, "Are you sitting down? Is my sister near you? I need you to leave from around her!!"

"She's not near me and I am fine. Just tell me what's going on…!" I tried to say something but he interrupted me.

"Mom! Momma, I've been shot…"

I had no time to scream or cry. This was not the time for that. I had to take off my mom hat and immediately think for the both of us. I didn't care what the reasons were behind any of this. I needed to focus on making sure he didn't bleed out and I no longer had a son. So I asked calmly, "Are you alone?"

"No, my friend is with me."

"Where have you been shot?"

"In my arm. I have a sweatshirt wrapped around it."

"Do either of you have a belt on?"

"Yes, I do."

"Take it off. Give the phone to your friend." I waited a second for the friend to take the phone. "Hello? Yes, take that belt from him and place it above the gunshot wound. Then I want you to tie it around his arm as tight as you can. I don't care if he screams because then I'll know it is tight enough. Plus, the bleeding will stop." I told him to be sure to secure the belt so it wouldn't loosen.

"Ahhhhhhh….!" my son screams.

I can't react was my paramount thought. I reminded myself that a screaming patient was a live patient. "Where are you guys at?"

"We had to run up the hill from where the car was."

"Is the shooter still around?"

"I don't know."

"I'm on my way, but I'll have to switch phones to find you. It'll take me some time to get to you so I'll have to call and get you help that's closer."

At that moment I was doing everything in a clinical and strategic fashion with no actual preparation. This was something I never thought I would be doing. However, right now, if I could sign a government contract where I agreed to be there for people that had no biological connection to me, then I surely could do this for my own child I gave life to. I made all the necessary calls and found a safe location to meet at. I couldn't release the phone line I was on with him and his friend because they might still be in harm's way. I stopped by a drugstore to buy medical supplies in the event I needed them once I saw him. God knew I had no time to break down and cry. I had to keep it all bottled in until maybe later.

Suddenly, it dawned on me that I needed to call in to work. I sent a text to my friend I worked with on the unit with the outgoing personality. She was the only one I knew who could handle a situation with details like this and not ask a million questions that I had no time to answer plus, help me under all that pressure. I sent the message. I had already tried calling the office from my phone and I didn't get an answer. I was trusting that she would get the message that I would not be into work due to an emergency. I didn't want to give them too much information because I was not even processing it all right now, and she understood that as well. She received the message and texted me. She assured me that she would handle my request with the greatest amount of confidentiality and I appreciated her kindness. Some people would be eager to spread the latest gossip, but she treated it as if it was a situation of her own. Even though working for the prison had its own flaws, it sometimes hired some genuine people that made the time inside bearable.

I made it through that gruesome and scary event with my son. I was trying to return to some form of normalcy. I didn't know if that was possible with all that was going on. I was able to let Hamza know about it and he was as panicked as I initially was. He was frustrated that he was not able to be at home during times like this when his family needed him the most, but what could we do other than work with the hand we were dealt. He called multiple times while we were all at the hospital. My mother was at my son's bedside. My daughter was not brought into all the turmoil but she did get a day off from school. In the hospital the phone reception was bad but I was able to connect and take Hamza's call in the parking lot. I immediately let him know that our son was alive. When messages were relayed through the prison sometimes important details got scrambled so I had to learn to tell him the important stuff first to ease his mind and take away his fear of hearing the worst. I then gave him as much of the details that I had gathered about the situation as possible. We had a little more to do, but soon my son would be released from the hospital and Hamza could call him as much as he needed to.

The stress of moments such as those had an effect over inmates as well. This was Hamza's second son who had gotten shot since he had been incarcerated. I knew this news would not be easy to deliver and definitely not easy for him to receive. I asked him to just stop when the pressure was weighing heavy on him and immediately pray. Unfortunately, with all of this going on, there could be no ideal or set times to pray. I told him to just do it and keep on doing it as he deemed necessary. I was still working through all of this and had not slowed down yet. I am functioning off of adrenaline alone, so I needed prayers on my behalf as well.

Well, time had stopped racing and everything was starting to settle down. I had a son whose arm was in a cast and who was nervous and ashamed. I can't process any of it. I am still cycling through my own feelings and the past 24 hours seemed to mesh together. I was just trying to keep my head attached to my shoulders and my feet moving one alongside the other. Once I sat down, I thought I would probably holler and cry, but that didn't happen at all. I was numb! I had to return to work the next day and I prepared myself for that as I got my uniform together. I hugged my daughter and let her know that things would be okay. We hadn't asked for any of it, but we would make it through.

I knew this was not the time to be critical of my son. He needed time to process what had happened to him and the role he played in what had taken place. So, I let him know I loved him and was grateful he didn't lose his life in the process. I could only hope that this experience was the one that would wake him up quickly and make him realize that life was short and could be taken away from a person in the blink of an eye. All of the things I had said, done and tried to instill in my son were all efforts to try and avoid something like this. But now here we all were, singing the music to a song no one wanted to dance to.

CHAPTER SEVENTEEN
Wedding with No Bells

All the preparation in the world couldn't prepare us for the day that was fast approaching. Marriage packets had been received with Hamza's notarized signature. I filled out my portions, letting the prison know that I was aware of who I was marrying. They asked questions such as, "What are the inmate's charges?" and "How long is his sentence?" They wanted to take the necessary precautions to make sure I wasn't being manipulated and lied to because it made for an unpleasant battle when trying to get out of any marriage. I had to go down to the courts and mail in certified copies showing proof of both of our divorces. I had even gone to the County Clerk's Office with another inmate's fiancé and the Imam who would marry us to turn in our marriage application. I received the official documentation that would be filed with the clerk after we were married in order to receive our marriage licenses.

The thing I seemed to be telling myself most often was, *"It is becoming real."* I thought it was becoming real as I ran around looking for dresses and making sure they fit properly, but this documentation process was making everything more relevant than ever. Just as I said in that visiting room three years ago when Hamza and I discussed our getting married, I told him then that I wanted the experience of a wedding, even though I was getting married in prison. We made a promise to each other that after he was released, we would repeat our vows on a date of my choosing by the water as had always been my dream.

For this wedding I was wearing a white wedding gown that I brought to the prison on my day off, to walk through the metal detector to guarantee it would not go off on my wedding day. This was not a normal preparation for any other bride-to-be, but a necessary step I had to consider if I wanted to have no glitches on our big day. I had to walk my dress through the metal detector on two separate occasions because the first time I figured out that the attached wristlet which would lift my train up off the floor as I walked had metal in it, so I had the unexpected last minute wardrobe alteration of making the wristlet detachable. A run to Wal-Mart for Velcro was a simple enough fix and a Google search for a seamstress in the area saved me from a bridal meltdown moment. And the second time I brought my dress through—minus the wristlet—to ensure the detector did not go off. All went well and the dress would make it through processing the morning of the wedding.

I had invitations made for family and friends to come and share in a celebratory meal after the wedding ceremony. Many of the people I knew would not be able to or had decided they did not want to go inside a prison to participate. For various reasons, legal or otherwise, I provided them a secondary way of being a part of our day. I catered food and reserved a hotel for all of us to gather after the wedding ceremony. There was an entire weekend of plans for me to attend to. I was not off on weekends and by sheer luck this entire weekend became available for me to request time off. Someone canceled their time off request and I was approved about two weeks prior to my wedding date. The rest of the month was already full on the calendar at work because only two employees can request to be off at a time. So this weekend was pretty much all or nothing for me. My thoughts are often filled with, *"So far so good, things are working out on our behalf."* I continued to pray that we could keep this momentum going!

As we had learned, nothing goes perfectly as planned. There was always a sudden glitch that must be dealt with and this time would be like all the others. My bridal party men and women were ready for the big day. His other children found various reasons to not participate, but my daughter and our son would be there. My daughter had to get specific approval from her father to come near a prison based upon our court orders from years ago. I have always respected his stance. So, I was surprised when he ultimately allowed her to make the decision to attend and he didn't put up a fuss. She and I were both pleased because I thought the outcome would turn out much differently. The only hiccup in all this planning was a major one.

I got a call a week before the day of the ceremony and was informed that the Imam officiating the wedding was going to be in Los Angeles, California at a conference and was unaware of the conflict in scheduling. Oh my goodness! I had not succumbed to a "bridezilla" moment until right now! What in the hell was I supposed to do? In order to have a wedding you needed someone to perform it no matter whether you were in or outside of prison. But that was not the part that was most frustrating. In a prison setting anyone coming into the prison had to be an approved visitor and listed on the inmate's visiting list. We are now "up the creek without a paddle!" I was trying not to kill the messenger as I was coming to grips with the fact that it was not the caller's fault and I had always known that "shit happens." However, this was a major one! It's not like I could just change the date and roll on with things. At that very moment everything seemed to be falling apart right before my eyes. Those envelopes I was sitting there addressing from the Facebook invite I'd already created; I was ready to toss into the trash. I wanted to just curl up in my bed and cry.

I called my bridesmaid and gave her the latest information instead. I know I was hysterical but she had been warned to be prepared when the shit started hitting the fan. So I can only hope she was ready for the flood of crazy I was about to unleash because I didn't know how to deal with this entire mess. It was someone's job to be rational and I had decided it was not my turn and not right now. She was trying to calm my nerves and suggested that I step away from the invitations, put them aside and told me to give it all over to God. We talked for a little while. Well, it was more like I screamed as she talked.

Then my expected call from Hamza came in as I was thinking to myself, *"Why am I always delivering this man bad news?"* I allowed the call to connect as I accepted the charges. He greeted me and I replied then gave a big sigh as I let him know we did not have an Officiant for the wedding because he had a conference to attend. He was silent for so long I didn't know if he had fallen on the floor or was just taking it all in. So I was calling out, "Hello?" to see if I would get a response.

He finally said, "Are you kidding?"

"Absolutely not!" I told him.

He was still stunned and I was sure his thoughts were like mine at this point; he was just not verbalizing them. He asked me a few questions and I was not able to answer them so I called my friend who gave me the information and brought her in on the call where he was able to ask the questions and get the answers he needed. Basically, the answers he was getting were still not changing the information we had received. My friend dropped off the call and he said in the most somber of tones, "It seems we always have to go through something. I can't talk to anyone in here at this point so I am going straight to my cell. I need to lockup to avoid anyone getting the brunt of what I am going through."

I agreed. It was just one of those things where doing nothing was the only option available. Soon our call was ending and we told each other we loved one another and ended the call with the greeting. Everything was so unexpected. I was trying to work through my own thoughts and it was not doing me any good. It was only driving me further into anger and disappointment. The only available option for me at this point was to just be still. The wind had truly been knocked out of me.

I still had to return to work and get through another day and I did just that. The sky was midnight blue and the moon was big and bright. It looked like it was directly over the prison as I was driving. I prayed and talked to God, asking Him to guide my footsteps and if this was just a bump in the road to let us maneuver over it with ease. I wondered if this was a sign that our timing was off. If so, then we had no choice but to follow God's guided path and halt all our plans. Tears were flowing from my eyes and I was so tired of taking the bumpy roads through life. I had to clean my face before going into that prison again and block out the part of my life that was turned upside down and present my best version of myself. These people had nothing to do with my personal life. I refused to allow anyone to see the struggles I was going through on my face.

I got to work and I managed to get through the day with ease, so much so that everything I needed to do was completed. I was sitting around with not much left to do except to remain in my head, thinking about the wedding that had no Imam to officiate. A whimsical thought crossed my mind as I was looking out a window on the unit that faced the parking lot and a field of trees. The thought was, *"Go and visit Hamza, now."* It was close to the end of my shift and I called down to the office and let them know I needed to leave early. There was no conflict with my request as there were about five other nurses on the unit just passing the time as their shifts ended soon as well. So I packed up and signed out and hurried around the corner to make sure I didn't miss the cut-off time.

As I was pulling into the parking lot I saw his sister heading toward her car. I got her attention and she let me know she had tried to visit, but something was making the metal detector go off, so she had to run to her car and make some adjustments to her clothing. I waited for her and eventually we made it in with no more issues. I let her know what had brought me around the corner to come and check on him. She was in shock and none of us knew what to do. Soon we were standing along the wall in the visiting room as Hamza came out. He saw me standing there and said, "I had a feeling you were coming." He gave me a hug then he hugged his sister. There were no seats at this point so we were standing and talking. I let him know that after last night's call and the mental state he was in, it hit me like a bolt of lightning today at work that I needed to come and see about you. Although I went to bed, praying and crying and woke up the same, I was okay. He assured me he was not okay, but would manage to push through. He just needed to go into his cell and pray, just as he did after our call ended.

The friend we talked to the previous night saw us in a huddle, talking and waiting for a table to be free so we could sit down, and she walked over to apologize again for all the changes and knew we were going through many thoughts and emotions. She offered a few suggestions, but none resolved the situation we were in. We listened to her advice, but there was nothing that would work for us. She left when we found a table to sit at. I could see the agitation on Hamza's face and I am sure my face looked the same. However, we made the best of the time we had together because we hadn't seen each other face-to-face in a long time. His sister's presence was comforting for us both. We needed encouragement so this wouldn't overwhelm us. Then as we continued to discuss things, we put on the table one idea about asking the visiting desk if they had pre-approved clergy that we could use to fill-in. Hamza and I were used to seeing a Hispanic lady performing ceremonies in the past. She was neither of our choices, but if we had to use her in order to keep things on schedule, it was a compromise we would grudgingly make.

As his sister and I were processing out, I spoke to the Officer at the desk and asked if they had a list of approved clergy. She told me there wasn't a written list, but the person who normally performed the marriages was visiting and she was sure she hadn't seen her leave yet. I was totally confused because the person who I had seen in the past performing marriages had not visited in months. And the man that I had seen occasionally, he didn't visit. He would just come in and perform the ceremony and leave. The Officer asked me to step aside and she would get my attention once she saw the clergy person. I did as she asked. His sister and I were talking and praying that this was the blessing we all had been asking for. Soon a lady I had seen multiple times was being steered in my direction. I never knew she was the person that performed the marriages. For the first time ever we introduced ourselves and she listened to my crazy story about having no one to officiate the wedding that was supposed to take place in a week.

Without hesitation she said to me, "If you are willing to accept my services and my fees, I have no one scheduled for Saturday. I will get the visiting form and I can have my husband and your fiancé connect so we can get the signatures and get the forms in to the staff by tomorrow. Then they can start processing the visiting form for me to be placed on his list of approved visitors, and even if there is a delay in the process I will visit my husband anyway and I can perform your service during our visit. No matter how we look at it, you will not have to cancel your wedding day. We can meet on Monday morning to go down to the courthouse and make any changes to your marriage certificate application if it is necessary. This is how God works! He comes through when we believe everything is going to fail."

For the first time since last night I am finally breathing again. I gave her the biggest hug ever. I was fighting back the tears that wanted to pour all over her shirt. Hamza's sister and I were weak with relief and knew we needed to share the information with Hamza. He was going to be so relieved, just as I was. As we walked to our cars in the parking lot we were giving God all the praise because we knew this was all a part of His plan. We had prayed through the storm and we would continue to pray right on into the victory!

When I reflected on the entire span of time from when I first met Vandrick at 14 years old to now marrying him at 39 years old, I saw how we had been through a journey that definitely was one that neither of us could have imagined, planned, or foretold. Our lives had been unbelievable! I am trusting God in all of this. We both have grown individually and now together. We have learned the value of communication, friendship, trust and respect. We have taken a spiritual journey together that we share between one another that has surely established within us a higher faith to tap into to keep us grounded. We did not place undue expectations on one another and it worked best for us. Neither of our lives had been easy or without trials, but we don't break easily under pressure. He had already served 16 years of incarceration in the California State prison system. I had done all I could to raise our son to the best of my ability while consulting him along the way which allowed him the opportunity to parent even from the confinement of prison.

I had made sacrifices that were out of my norm as I went through my own version of incarceration when I committed myself and family to the military. I was challenged to live without many luxuries; trained to tap into survival skills while fighting through sleep deprivation; trained to fight to the death if the situation presented itself; equipped to thrive off the land and use nature and minimal supplies to guide my path. I had to depend on complete strangers who would later become like surrogate family members. I had to adjust when family members who told me they loved me had proven even they would not always remain loyal, tossing my secrets into the wind as I learned they took private matters and made them public. But I also realized comfort was all mental when you're tired enough to sleep anywhere. Hamza and my lives have had twin aspects where we can now review, share and gain an unspoken level of respect for one another that many closest to us will never understand. But it was okay because the only two required to understand it were us. I was just days away from giving Hamza my life for all of eternity. We would transition from friends-with-benefits to officially husband and wife. We would now become one as I joined with him fully into our incarcerated life.

Our wedding day had come and I didn't sleep much at all. I was up until 2:00 a.m., making sure everything was in order and everyone was ready for the next day to arrive.

It was just us girls in various hotel rooms and I was in a two-bedroom suite where my sister was staying with me. She had stated from the beginning that she wouldn't go to the prison, but she was here to make sure things were together at the hotel, including ensuring the food arrived and was set up by the caterer. My cousin, who traveled from Los Angeles to be one of my matrons of honor— I was blessed with two—was in the room on the other side of the living room with my daughter who was my flower girl, although we were not allowed to bring flowers. His sister was my other matron of honor along with my bridesmaid, a friend who was engaged to be married to one of Hamza's Muslim brother, were next door to each other one floor down. Everyone I wanted to attend couldn't make it which I expected, but those that were here, I am grateful to them for sharing their time with me.

My mother had gotten sick, but was insistent on still coming so that she could celebrate with us after the wedding. And she even promised to be on her best behavior with my dad being present and not reflecting on their previous experience. I saw her the day prior when she gave me a monetary gift as I made it to my hairdresser. My fabulous and very talented niece would arrive at the hotel by 4:00 a.m. to flawlessly apply everybody's makeup. I chose her because she got the look just as I wanted without any errors. Before I selected her I went to Macy's Department Store to the MAC cosmetic counter for a trial session and was highly disappointed as I walked out, feeling like a circus clown with all the make-up that was plastered over my face. My dress hung on the back of the double doors of my room and I was almost ready to pour it over me and have each button hooked. After the prompt 4:00 a.m. appointment time, I took the first glimpse of my make-up. Oh my! I wanted to shed sentimental tears at the reflection of beauty smiling back at me. But everyone ran over and reminded me, "Aww… but you can't mess up your face."

Luckily my niece was still ready to work as she laughed and said, "You can cry as long as I'm here. After that, no more tears." She was paid generously for her services and told me she needed to get back on the road because she needed her rest from a long night and early morning But I respect her hustle as "Milly Made Me."

I wanted to be on the road and headed to the prison by 7:30 a.m. We had to stop by my house to pick up the best man, our son, who was awaiting our arrival. We all loaded into my seven-passenger SUV and got on the road.

As we entered the prison grounds it felt different than any other time. I am in the passenger seat and reciting the Arabic prayer, Al-Fatiha, which I learned as a surprise for his 41st birthday almost a year ago with the help of his former "celly," friend and Muslim Brother. We found a parking space near the entrance where we would process in. The Officiant was there. The visitors waiting to be processed in were in awe when they saw me in my wedding dress and the rest of us dressed in gowns with our son wearing traditional Muslim attire. We definitely had everyone's attention. What I had planned and prepared came together so beautifully, even impressing myself.

I was experiencing a few jitters at this point, but my matron of honors and bridesmaid knew exactly what to do to keep me from losing my cool. I had warned them many times that I might lose all of my cool points on this day. So they were prepared in case of an emergency. I told them to catch me if I started to fall and remind me to do the simple things like breathe. Next, we all processed in and then went downstairs to the bathroom to use their mirrors to readjust everything that had to be removed and replaced to get through the metal detectors, like the hijab the three women were wearing. Once through the detectors I waited on the sofa for Hamza to enter the visiting room. We had a long table reserved so we could accommodate everyone in our party. He could also have three inmates to come out with him and two of them were our witnesses. Well, one witness was with him and we were waiting on the other one when we received an unexpected message from the Officers that he had gotten into a pissing match with some CO's in the back and he would not be able to join us.

By this time, the visiting room was receiving its normal visitors and we had many inmates coming to circle around what was now our official ceremony. We were beautifully serenaded by Hamza's sister who sang a verse and chorus of the song chosen by Hamza, "Here and Now," by Luther Vandross. The Officiant performed the service with such ease and we both were filled with emotions as we shared in the realization that through the will of God we were able to have this moment in time together. Afterwards, Amir performed a second ceremony that was representative of Islam. We took our Nikah vows which we had prepared and recited them to one another and shared with friends and family. They were our promises of consistent efforts toward our relationship and our commitment to one another and our family. As we came to the end of our vows we are led to recite in Arabic, for those who know it, the Muslim prayer Al-Fatiha. Then another brother recited a prayer that he selected for our union as Hamza and I stood and received it under the sound of his voice. Our guests remained circled around us and the photographer stood on chairs at times to capture every image that he could. We took many pictures, during the ceremony and afterwards. Then we headed back to our long table where we were greeted by many handshakes and hugs. The people in the visiting room made their way in our direction to compliment all of us on a ceremony they had never witnessed before. They could see that such great thought and effort had taken place to make this day special despite the conditions afforded in a prison facility. Even the staff and CO's that normally refuse to give no more than a grimace at the visitors, kindly said, "I can truly feel the love in the air. This is definitely not a normal prison marriage." And they even wished us well.

I appreciate the acknowledgement and it was true, I didn't want to mirror what I had seen of marriages throughout the prison system. I was not taking anything away from their weddings; I just hoped they were all as fulfilling as mine. I knew I was giving Hamza the full and complete me! That decision had not come easily and I didn't want our ceremony to be dampened down because of his location. One thing was for certain, after seeing the outcome of this moment, it can be repeated to those not present to the best of our ability, but what spiritually happened inside this visiting room could never be duplicated. This moment was surreal, even for us. The journey Hamza and I had taken had been so far from being traditional, unplanned, unforced, but it had never lacked excitement. Everything about us had been a journey of unexpected turns, endless possibilities and solely in God's plan and timing.

CHAPTER EIGHTEEN

Hunting the Good Stuff

We captured all of this in our ceremony. The marriage license was signed and the Officiant took it to her car for safekeeping. She would return in a little while where she would re-check in and visit with her husband uninterrupted. There was no way Hamza and I couldn't acknowledge God's hand in the midst of it all; how he and I went from one ceremony that was supposed to be performed by an Imam and it falling through, to us participating in two ceremonies that uniquely represented us both. God truly wanted us to know that He was in control of it all. He had ordained this union and it would be abundant and fruitful where we both would flourish under his spiritual light. Hamza and I are amazed! This was definitely not our own doing. It was more than either of us could have ever imagined. It was a miracle from God!

We had enjoyed the most loving and memorable ceremony ever and now it was time for the reception dinner at the hotel. We had guests arriving in just a few hours. The day had already surpassed any dream I could have had about it, but I knew I had to keep pushing with the same energy because family and friends would be festive and full of congratulations. Before they were scheduled to arrive, the bridal party—minus my kids—had a quick, impromptu session of photography in the boudoir of my hotel room that was just for the groom's eyes as a honeymoon gift he could enjoy in a few days via the mail. We might not be able to consummate our marriage immediately, but that won't stop much. A relationship, such as this one, required imagination and creativity. I could handle the challenge and I had just enough support around me to make it happen when needed. They were reliable, unbiased helpers who came in handy at times such as this to help me pull off the best gift ideas.

My dad was the first guest to arrive at the reception and I was happy to see him. He had originally planned to be at the facility with us, but during the paperwork portion he became reluctant to complete it because he was worried about going inside a prison. Not that he lived his life in a manner that would cause him conflict, but it was the fear of the unknown or the possibility of something setting off a trickle-down event that had him shying away from coming inside. I was initially upset, but later I came to understand his position and I got over my feelings. I know my dad's love for me was strong and he would be there for me on the outside. My sister and mother had already adamantly stated they were not stepping foot in a prison, and they both meant it. Even if I was an inmate myself, they were not going to change their position.

My sister had already told me, "I will be with you through it all until you're walking out the door," and she did just as she'd said, with hugs and kisses as we left the hotel.

My mother just kept reminding me, "You will have to do this all again for me when he comes home. I want to see my baby girl get married for myself."

I assured her that it was in our plans to have a second ceremony, not so much for everyone that didn't attend, but for the parts that we couldn't substitute in the confines of a prison because Hamza and I both loved the tranquility of water settings. Therefore, we could enjoy the ambiance of our family, hopefully in their entirety, the next time around. If they were willing and able to travel, our hearts desire was to have a "destination" wedding ceremony.

My mother, sister, brother-in-law, aunts, cousins, nephews and friends have all come and made a lasting impression through their pictures, the traditional toast, the meal and the cake. Hamza even called while the guests were still there. They shouted through the phone in unison, "CONGRATULATIONS!" as I stood, grinning from ear-to-ear. I was certain he was feeling their excitement and well wishes on his end.

I could hear the smile in his voice as he responded, "Thank you all!" over the speakerphone.

I took the rest of the call in my room so I could hear him better. He was sharing with me the many compliments from the guys on the yard he saw in passing with remarks such as, "Hamza, you and your wife did that! We have never seen anything like that." He told me one guy even mentioned that he walked out of the building early in the morning and looked toward the parking lot and could see a wedding dress in the far distance and wondered who would come to a prison dressed like that? He later heard all the talk about Hamza's wedding, the bride and the bridal party and realized that the wedding had to be for his Muslim brother. I was happy to hear him say the inmates in his building prepared a potluck meal as well as their own way of sharing together in their own reception dinner for Hamza. It felt good to know they were so thoughtful and that he was being celebrated on his side as well. His call would end in its normal 15-minute fashion, but we ended it with an added sentiment this time.

I said, "I love you, husband."

He replied, "I love you too, wife," with a chuckle as we both thought, *"How did we get all the way over here?"*

Our guests received favors for coming and even signed our wedding albums that I had pre-made a few weeks prior just for this moment. I had two albums made. One was of our now joined family members. All of our children are in various photos with Hamza or with me; some of them were by themselves, a page filled with a collage of pictures, and there was no way that there was not a tribute page to his mother and grandmother who proceeded this moment in death as well as a page dedicated to my mother and father. This album portrayed where we came from and the extensions of who we were. Then there was an album made just for Hamza and I which showed the many years that we had grown up together. It shocked everyone when they saw the pictures of us as young teenagers and then the pictures of us now. I even had an enlarged photo of us a couple of months before he was taken to jail, holding our son in his arms. These memories, and now the signatures and messages from our guests who support us in love, we would have forever to reminisce over throughout the years to come.

It was getting late into the evening and the day's events were wearing on me and I was getting sleepy. I was supposed to do a wardrobe change, but this wedding gown was still on and would be on until I called it a night. I went downstairs one last time to arrange for my now, sister-in-law, to stay an additional night as it had been a long day and I didn't want her traveling because it was late and we were all tired with so few hours of rest under our belts. I also let the hotel know I wasn't accepting any more guests. I was about to call it a night and those not staying the night would be leaving shortly. The hotel staff was so nice and had enjoyed sharing in my entire wedding experience. They had been hospitable and fun throughout our stay. It was a great idea to plan on having the reception here as well as staying here versus the original plan of everyone going to a restaurant and renting the large banquet room. I liked the intimacy of the hotel experience and our caterers provided timely and professional service as well. Today had been far more than I ever planned in my head or envisioned in my dreams. I was a happy bride and had proven that with prayer, everything would come together beautifully.

I returned to my hotel room and lay across my bed with the double doors open. I was so grateful that many of the people remaining were family because I was tired and I didn't feel bad that I was not still on my feet, entertaining. They were free to take whatever food they wanted home. I was in no way going to be able to eat it all. My sister and cousin, at least it sounded like the two of them, were in the suite's kitchenette, tidying up a bit. My mom was coming down with a bad cold so she had been in the second bedroom most of the evening. But, I heard her voice as she was getting ready to travel back home after dropping off one of my aunties who rode in with her here. My phone chimed, indicating a voicemail had come in on my cell phone. I listened to it and it was one of my current co-workers who said she had arrived and had called me from downstairs. She had not said where she was downstairs, but it was late and I was not expecting any more guests at this hour. The message was left at least an hour or so ago so I assumed she was not still waiting to come upstairs so I figured she had left, especially since everyone, including a co-worker I worked with years ago, made it upstairs to the reception location. Plus, the hotel staff did a great job of providing information to our guests if they were unsure where to go.

I didn't let it consume me because I was definitely readying myself for bed. I started smiling at the memories playing in my mind while my eyes were trying to close as sleep was getting the best of me. Plus, I was scheduled to visit my husband, Hamza, in the morning. I felt bad that my co-worker had driven here and still missed out on all the festivities, but I wasn't really expecting her to come. I thought it was more of a joke when she and another co-worker kept insisting they were coming. I gave them both two of the extra invitations that remained after I prepared them for all of my family and friends, not really believing they would show up, but more as a souvenir from the wedding. I knew it was a joyous time and experience, but I was unfamiliar with people inviting themselves and actually attending. Under normal circumstances, weddings were prepared with headcounts and RSVP's months in advance along with a strict guest list to maintain a budget. I hadn't known either co-worker very long as I was still considered a new employee. I appreciated their warm welcome and acceptance of me, but neither of them knew the abnormalities behind the planning of my particular wedding and marriage to an inmate. I knew nothing about their private lives. I remained consistent with my rule of not freely sharing my personal business in my professional environment.

As the hotel quieted and after all my guests had texted me to let me know they made it safely to their destinations, I rested with the biggest smile on my face, relaxing now that all the fuss had come to an end with the anticipation of another visit with Hamza in the morning. I looked forward to this visit with baited breath, knowing it was bittersweet. I would get to spend the day with him and we could reminisce over everything, but also realizing that this visit would be the last one for a long time because I still did not have weekends off. I got off too late to make it around the corner with enough time to get on the facility grounds before they stopped visitors from entering. This was hard to face, but it was our reality. I would do my best to keep this thought as far in the back of my mind as I could and not allow our visit to be overshadowed and instead remember all the great things worth smiling about. We deserved to have as much of this time together filled with joy as humanly possible. We won't get much of a honeymoon. So as we are learning, especially me, "the experience is all in what you make of it." The smile I have is the one that I created. I would make the most of the photographer's backdrop with the beach setting. The photographer, an inmate himself, made an effort to take the best pictures. I always remind him that JC Penney's would be crazy not to hire him once he was released. He did all the similar things their photographers did like snapping and making noises to get little babies' attention for the best photo; made sure the lighting was just right; positioned families for the placement of the backdrop they'd selected; and allowed the family to check out the proofs before they were sent to print. He did his best to make sure his customer base was satisfied with their photos, even though they had to be approved by the CO staff at the podium.

The next day, my visit with Hamza was routine, but full of fun nonetheless. We spent much of it reminiscing over the wedding day and evening before. We wished we could have shared the entire day together, but our individual events were well planned out, so we enjoyed every one we were able to spend time with both together as well as apart. We ate our normal vending machine meals together, and spoke of our future plans as they seemed a bit more realistic now. All the plans of the past were our reality now. We were one unit, one force and definitely in this together. I felt an increased passion to fight to help Hamza get out of this place. I knew not everyone had a chance and shot at coming home. Some criminals had earned their time to be in this place. I was far from naïve and understood the establishment and necessity of the justice system. I had faced my own situation of being a victim of crimes committed, but I also strongly believed that a person should be convicted properly and receive a sentence which fit the crime they committed. The penalties should be justifiable and they needed to discontinue this system of inequality with trumped up charges; systems punishing people based on trying to make a name or title for themselves; enough of the craziness that boldly echoes the ideas of, "if you have status and money you are able to avoid the penalties of your crime, while the poor and under-educated will never see the light of day."

I knew I was not certified with the degrees of knowledge that an attorney possessed. However, that would not stop or undermine my ability to work diligently at fighting for what was right. Honestly, the way I looked at it, I could be more effective not having all the responsibilities of being an attorney because I would not be burdened with this being my daily responsibility and obligation or have my unique opinions and methods countering any legal stances an attorney must take. I knew attorneys sometimes had to move cautiously with their strategies based upon the opinions of their peers and legality of the laws. I would leave all of that up to them as I performed the background investigation and put the knowledge I did have to work as I tried to pry my husband out of this institution.

I was able to enjoy two more days off from work to get all the decorations put away and my normal paperwork and tasks I usually took care of before I returned to my regular work schedule of long and unpredictable hours locked away in the prison. This time, going back as a married woman, I felt a bit different, especially as my co-workers had heard I was getting married and many of them wanted details, but I remained as discreet as I could without hurting anyone's feelings. I was determined to keep the professional boundaries that I felt most comfortable with. I had one co-worker and her husband who didn't make it to the reception because she was sick. She was so disappointed that she couldn't make it and I hoped by the time I returned to work she was back and feeling better. I would make sure to bring her pictures of what she didn't get to see as well as the lady on my unit who had been a great help and encouragement along the way, especially on those rough days. I knew they couldn't make it because of their own reasons and circumstances, and I completely understood, so I hoped the pictures would suffice.

My return was pretty much the normal signing in, grabbing keys and heading up to my unit. Everyone was there in their normal jobs: breakfast trays were being issued, nurses were sitting with staff that needed additional attention for various reasons, and I was trying to find out if I would need to distribute medications, or prepare to work the floor. The other nurses were trickling in and receiving their reports of the prior night events. Everything seemed to be a normal day with not much of a fuss. I settled right in; medication distribution was for me. My co-worker worked many hours of her regular shift and a ton of overtime hours so she would not have to pass the meds this shift and instead sit with a patient one-on-one. I truly understood and had no issues with the request. She worked a crazy amount of hours on this unit and I was fully rejuvenated so being up and about didn't bother me one bit. She and I worked well together in our ability to accommodate each other's needs. Plus, because there were established breaks when you pass out meds, I would also have a time set where I could go downstairs and check to see if my co-worker friend had returned from being out sick. I knew where she tended to work most of the time.

My break time came and I headed to check on my co-worker friend. I hadn't heard from or seen her yet today. That was normal as we tend to sign in and head straight to our work locations. I'm hoping she was there and back to normal. She was such a fun-loving person and for me her personality was infectious. If I was having a bad day she knew just what to say to get me laughing. She had been working here a number of years so this environment was far from a challenge for her to work in. She told me she had learned to roll with the punches as they came. She, among others, had told me one year is about the general timeframe that it would take for me to become used to working here. They all said, "Give it a year and the craziness will bother you less." I rolled my eyes with a simple, "Yeah, yeah." I knew I was not accustomed to adopting a total disregard for organization and structure, so working in this prison environment was having a definite effect on me. Safety was a priority I just couldn't turn aside.

I headed down to the area she normally was in and asked the lady whom I received the voicemail message from if she knew where our co-worker was?

She interrupted my question with a, “Oooh, I am mad at you.”

I gave an apologetic smile and replied, “Good morning and I am truly sorry. I received your call but by the time I got it I was getting ready to go to bed. Also, why didn’t you just follow the people who were heading towards the reception? I had people in and out all evening. That’s how a few guests honestly found their way.” I knew exactly what she meant when she said she was upset. I went and sat in a chair in her area after giving her a hug. After finishing up the last of what she was doing she came and sat next to me.

“I brought a guest and we waited a while,” she said. “I saw lots of people, but didn’t think to follow them.”

I listened, but I wasn’t sure how I could fix things at this point. There were no do-overs. So I apologized again. I was curious to know what time she arrived since I got the message at almost 11:00 p.m. So I asked, “Why did you come so late? The reception was scheduled from 5:00 p.m. until 10:00 p.m.”

She informed me, “We arrived about 6:00 p.m. and the hotel was unsure where to direct us.”

I really felt bad and confused as to why I hadn’t received the message until so late at night. I had my phone with me most of the evening and there were only a couple of times I put it down, but it wasn’t on silent. This bothered me a bit, but again I couldn’t fix it. I could only apologize that she wasn’t able to attend. Since I had the pictures of the wedding in my jacket pocket for the other co-worker friend to see, I decided to show her the pictures as a consolation for her not being able to attend. I did feel bad that she drove all that way to attend and didn’t come upstairs, mingle and enjoy the food and my family that were also there to celebrate my union. A bit hesitantly I showed her the pictures because I had given her no warning that she would see a wedding in a prison. As she looked through them, based on her initial impression, I knew I would have nothing to explain. All prisoners wear the same type of clothing, so the inmates in attendance were no different than what she saw everyday so there was nothing to explain. I noticed she stopped at the first picture and gazed at it a bit then looked up at me. I smiled at her surprised expression.

She looked down and said in a faint whisper, "Oh, I didn't know.... you look beautiful."

I nodded and said, "Thank you."

She continued to thumb through all the pictures until she reached the end. She didn't say much else. She admired everybody's appearance and acknowledged that we were all beautiful. I appreciated the compliments. My friend and former military co-worker came in behind me and I was glad to see him. He lightened up the mood and a room no matter where he ended up. I happily showed him the pictures and he gave his normal joyous responses and congratulatory acknowledgements. He knew my background story the longest so I knew none of this was shocking for him. He was just happy to see we finally were able to complete the ceremony and was pleased to see my son was in attendance. He knew about the dilemmas I have had to face with him over the last few months. After the two of them finished with the pictures and compliments I headed back to my unit. I couldn't stay away too long. Patients could be requesting medications so I had to get back because I was the only person with access.

I headed back to my unit and thankfully, no one had requested any additional meds. All is well on the unit. We had a pretty routine bunch. Once they got their morning meds they didn't require too much more, but it was always good to be available just in case there was a change in their medical status. I still had time to prepare myself for the afternoon meds and to check on my diabetic patients. My co-workers had even given each other breaks in my absence so there were no issues to return to. Good! I wasn't gone long, but in the medical field anything could literally turn upside down in the blink of an eye.

I am out on the unit and I received a call right as I was completing my mid-day insulin injections. I walked over to the phone and what I thought it would be, was the manager's office calling with their daily checks to see if anyone was available to work over-time. The first call was always the volunteer request and the next call was the mandatory call. We all received an email daily to see what number we were on the list for mandatory overtime. This was the dreadful list that guaranteed you would be working a 16-hour shift if you were close to the top. On occasion it had not mattered what number you were because they had no staff showing up for the next shift so almost everyone got hit. This was what I was mentally prepared for this call to be, but this male voice was far from familiar.

"Is this Nurse Riley?" he asked.

I replied, "Yes."

He continued, "This is an Officer from the Investigation Department. Did you get married this weekend to Inmate Jones over at Solano?"

"Yes, I did," I responded with curiosity in my voice. I am wondering why in the hell I was getting this call and on my unit of all places, but I would go along with it since I was already in it.

"Well, I'm calling because I want to make sure you have completed all the necessary documents and this information is in your file."

"Yes, I did. I completed the documentation during the hiring process," I told him.

He replied, "Oh, well that paperwork is normally not the appropriate form. We have a more official form here in this office. Can you come to our office and complete it?"

"No problem. I can come now," I said.

"Do you know where to go?"

"No, I don't."

"Oh, well that's a good thing," he said with a small laugh. "That means you've never had any issues here."

"No, I haven't," I responded, less amused than apparently he was. He gave me directions and I excused myself from the unit and left my keys with another nurse in case this took longer than I expected it to. I followed his directions and soon I was knocking on a door in a seemingly hidden stairwell. Someone answered and I entered the office and I was looking at one officer in front of me and one slightly down the hall standing in a corner. I don't know if he saw me or not, but I was told I could have a seat at the table by the Officer who introduced himself and let me know he was the Officer I just spoke with. He let me know the department had received information about my marriage and they just wanted to make sure they had the appropriate documentation.

I listened, but I was far from slow as I knew this was necessary information to find out if I met the person referred to as "Inmate Jones" here because that was immediate grounds for being walked out, pending further investigation. Staff and inmates cannot engage in improper relationships, but I would play along with the song and dance. He brought me a form and it was more detailed than the one that I received during hiring. They should have attached this form with the hiring application to avoid any inconveniences such as this one, but I knew I was actually asking for there to be a perfect world and that would be asking for the impossible in here. As I was filling out the documents, the Officer started fishing a bit. I would give him just enough information to let him know I was no fool.

I replied, "We have a 20-year-old son."

I could see he was caught a bit off guard because he was expecting something more underhanded to be taking place, so he remarked, "Oh, you've known Jones a long time?"

I told him, "Yes, since I was 14 years old. So no, I didn't meet him as an inmate."

He stumbled to catch up with his thoughts as he realized he hadn't caught a fish on the hook this time. He said, "Oh, in that case, you can just put that you and he have a son on the back side of the form. You can write that you've known Jones since you were younger. We'll put it in your file and make sure the Warden has the information so he's aware."

I finished up with a brief summary on the back and handed him the document as he walked me to the door. We said our goodbyes and I headed back to complete my shift. A million things were racing through my mind as I was walking back to my unit, but I refused to let anyone ever see me sweat. So I put on my "happy go lucky" outward appearance and let what just happened sit in the back of my mind as I pushed through another day. I was so grateful that I took care of myself in the beginning or this would have turned out so differently. I'm glad I filled out my application documentation properly and asked all the right questions before I accepted the job offer to enter into a prison facility. Most people thought they could slip through the cracks, but they quickly found out that people talked too much and couldn't wait to spread gossip; and the person they least expected would be the one to unexpectedly cost them their job. I said a simple prayer of gratitude to God for His grace and mercy and entered my unit to finish out the remainder of my day.

After the experience I'd just gone through, I had thoughts circulating in my mind and I kept asking myself, *"Why, almost instantly after talking to one of my co-workers, I was called on the carpet about my marriage, knowing the significance it could hold over my continued employment?"* I was glad I wasn't tangled up in the names of mandatory over-time sweeps. I needed a break to clear my head and a call from my husband would help, and seeing him would be even better, but that was not the case so our daily phone call would have to do. I received my normal collect call shortly after making it to my car. I was trying to keep my composure as I didn't want to hit him with an immediate, "Guess what happened today?" That was definitely not a great way to start any call, unless it was absolutely unavoidable. That opening was only used for a 9-1-1 emergency. This call was more like a 2-1-1 resource information call. So I would take it easy as I accepted the collect call.

We greeted each other with the Muslim greeting of peace and blessings and he started the conversation by telling me about his day. He was enjoying all the goodwill from people acknowledging how beautiful our wedding was and how it had shed some light upon prisoners who often lived in darkness from the daily grind of incarceration. The wedding had brought a glimmer of light, even if for a moment. The Imam employed with the prison had heard about it as well and congratulated Hamza on being able to pull it off, especially since he was our initial choice to marry us. We never took it personally that he could not marry us due to a conflict because we understood the potential risk and legal ramifications it placed on him with regards to his religious and personal beliefs if he did not choose to provide the service to everyone that was eligible to be married. We remained optimistic that we would eventually find someone to perform the ceremony and as everybody could see, things worked out just as they were intended to. I was elated to hear Hamza was still in a beautiful mood and all was well on his end. Despite the situation I had just undergone, I was in a good mood as well. I was still reveling over the fact that we had an incredible wedding and were on the path to a fulfilling marriage.

However, the reality of being called to the Investigation Department, even though there was no merit to their suspicions about an improper relationship, still didn't sit well with me. The fact that it would have been a display of professionalism by doing the research necessary before making a hasty judgment call. Had the Officer taken a short walk to the Personnel Department or with minimal effort sent an email to check my personnel file, they could have avoided meeting with me altogether. The information requested had already been provided during my hiring process and if it was not included, then they had sufficient grounds to question me and my intentions with regards to the concerns of safety of their facility. However, as I have witnessed more often than none, working here is like flying by the seat of your pants and is unfortunately an acceptable practice of dysfunction.

I let Hamza know what I had experienced a few hours prior to his call. He listened attentively, but I could tell he wasn't too fond of the topic. I gave him the time he needed to absorb all of the information. I told him my speculations about what took place and why I think it happened. After he listened he drew the very similar conclusion that I had drawn, but was trying hard to fight off his suspicions. But, coincidences such as these just didn't happen without a bit of encouragement, and we were both certain of that. I had been visiting CSP-Solano far too long and too many times as an employee for there to be this sudden inquiry about who I was visiting, especially since some of the Solano staff had known my employment location and occupation long enough that if there were any concerns it would have been addressed way before I ever got married. We both worked through the situation, trying to find the positive side, "hunting the good stuff," as my military training would encourage me to do.

But, it was difficult as we both got sick and tired of the constant ridicule that was associated with families, relationships, and bonds with inmates. We were able to do what we had learned to do over all these years and not let the messiness and stigmatisms of prison relationships and prison lifestyle get us down. We reminded each other that one day this sentence would be over. And what in the hell would the rumor-mills, gossipers and shit-starters have to talk about then? So we continued to do whatever the heck we wanted to do right in their faces without a single care in the world. The irony of it all was we were doing just that right now, living our happily-ever-after with far less than most have access to daily; and while folks are not moving forward in their lives it's likely they were too busy speculating, watching and talking about lives that had no direct effect on their own.

CHAPTER NINETEEN
Rules of Engagement

I now wake up every day, smiling and thanking God, because I was so content at the decision I made to accept this man's hand in marriage. I hated that we couldn't spend every waking moment together, but I knew that time was soon approaching. Even though we were not discussing it publicly, we were working together and had built a team with the help of one of my matrons of honor and her husband who provided us with a necessary contact that believed in Hamza's innocence and could help us to fight his case properly. The facts of his case could not be disputed: Hamza was not the person behind the gun who pulled the trigger that caused the victim's bodily harm. The system would have to right its wrongs. No person should be charged with a life sentence because they were choosing not to collude with detectives, prosecutors and defense, especially when there was scientific evidence readily available for the case. Evidence was being tampered with, illegally destroyed, and hidden from trial in order to assist the prosecution in railroading someone in order to slap a closed sticker on the case then moving on as if the crime they were committing in framing people was not as unethical as that of any hardened criminal. I was always led to believe the judicial system wanted to find justice for acts of violence and crimes against citizens and not just close cases, but I quickly found out how wrong and naïve I really was.

I have invested in pulling his entire transcript from the Court of Appeals located in San Francisco, California. I would break it down and prepare additional copies for our legal team. Yes, I was running around just as much as the team. From the way I looked at it, and always will, no one was going to place more urgency on your situation than you are willing to place on it yourself. Hamza and I had discussed his case in the past, but I had never read much into it until now. Yes, I had seen the newspaper articles that painted him as this big gangster to reckon with like the famous names of John Gotti, Al Capone, Felix Mitchell and many others. I'm sure those who knew him personally would agree Hamza, more readily known as "Boo" at that time, was active in the streets and doing what people in the drug business did, but I was certain all could agree, he hadn't made it enough to be considered as big a name as the newspapers would have its readers to believe. There was a bit too much theatrics thrown in, but so was the trial.

I read the articles with the same intensity as I had read any other criminal case. I knew that journalists must build up a story to draw in its readers. They must sensationalize much of it for their own editors to take an interest so it would make the cut and not end up on the floor with all the other less desired news. So I gave the journalists credit for doing just that as the headline to an article boldly read in the California Oakland Tribune, "Drug dealer faces 40 years in prison." It appeared in the April 18, 2004 edition of the local news section by Glenn Chapman, Staff Writer. It read, "A self-described crime world 'lieutenant' found guilty of trying to execute a small-time marijuana dealer during a drug rip-off may spend the rest of his days in prison." So, from suggestive editorials and a case that had already reached a verdict we would have lots of work fighting through the post-conviction remedies to make sure that the right information was finally revealed.

Hamza and I spent parts of our time on the phone, discussing much of the case and after so many years some of the details were getting lost in translation, but he was definitely sharp on the main points that his Public Defender didn't investigate or openly told him that it was a waste of time and money to follow up on physical evidence that was readily available. As we talked it was hard for me to hear some things because I still had that little girl passion in me to advocate for people's rights. So, I was floored when I was a witness to individuals who were hired and paid to perform a duty where they spoke about how highly they honored their work ethics or how they took an oath and swore to uphold it in order to gain employment, during the interview phases. But once they've gotten the positions they were the first to find every shortcut possible to avoid what was expected of them. It was not a practice I understood nor would I allow it to be justified.

The most important and intimate emotion I tapped into was the sound of my husband's voice, cracking, as he spoke of the gratitude and appreciation that he had for me in doing what I was willing to do for him. He was in awe of the amount of time and dedication I was willing to put into his case and fighting for his freedom, not because it was a request from him placed upon me as my obligation as his wife, but because I gave a damn about him. It was proof of how much I cherished our marriage and my desire to help my husband to come home where he deserved to be. The place I knew his incarceration would have the greatest effect on was his own family as he tried desperately to keep them away from the life he once lived. He told me, "You have done more for me, legally, in these past three-years that we have been together than anyone else in the total sixteen years I have been incarcerated. You have come on board, writing and talking to paralegals, lawyers, Inspector Generals and government officials as if you are fighting for yourself. You have taken things on as strongly as I have in the past and you are not afraid to step in. I will never be able to thank you enough for what you have already done."

I appreciated his sincerity because I had come to see and understand from being employed by the same entities that confined my own husband how they operated. It was a daily task to wake up every day and face the many men and women who are locked up and the keys thrown away. They had lost their bonds with family and friends who had gone on with their lives and forgotten about them behind those walls. Some were locked up for crimes that they weren't even responsible for. Trying to fight through the red tape of the system had caused many to just give up hope, faith and the belief that they could overcome the obstacles they faced. So, they wait silently for a door to unexpectedly open, perhaps their name and number would be called, hoping that someone would recognize the errors and falsehoods, or that someone would eventually give a damn.

Hamza and I had laced up our gloves on so many different issues that I had to build a full filing system at home. One was for the case he was incarcerated for, the second was for violations that were committed by the facility and its staff, and the third was for the continuous changes to the law. However, I could only keep things that pertained to him specifically. Even though I had friends and family that also needed information from time to time, I found out some people would not fight for themselves but would sit and wait for me to act on their behalf. I was a kind-hearted individual who did not mind helping others in need, but I did hold to a golden rule: I was only willing to help anyone to the level that they were willing to help themselves. I had found out over the years that people would go on vacation while you worked everything out for them. I could give information and leads all day long, but running out and researching, faxing, emailing, typing and everything else costs. I had a fee schedule for that and my work did not come cheap. It was a known fact that I did not sleep like normal people, but I dealt with sleep deprivation based upon my choice and completing things beneficial to me. So if I were to handle someone else's work, then I would be paid for my time and efforts and I guaranteed the proficiency of my return for the request. From word of mouth alone because of the amount of knowledge I had gained, I had been asked to consult with families as they were adjusting to this prison lifestyle. It's always a sincere pleasure to provide my services. The part that brought me the greatest amount of pleasure was witnessing them execute the plans that I had provided then hearing their success stories. The joy in the small and large victories is just enough reward for me to know that the work is worth it.

My biggest and main suggestion to a prison wife, mother, sister, girlfriend, or associate is to remember that there was power in numbers. Don't be afraid to speak up if they noticed something was being done wrong. I would tell them if you or your spouse's rights were being violated and you could prove it, then it likely was not just happening to you. Someone else was likely going through the same thing as you and your loved one. So, if you can find someone else going through it, do your best to bind together and fight against what was going wrong. There were a few established organizations that fight for the rights of inmates. The few that existed were overwhelmed and often couldn't take on every case. However, if it was something they knew would affect the masses, they would try to help take action on their findings. The government was finally starting to place some importance on prison reform and mass incarceration campaigns. It was helping smaller issues get the attention after having been overlooked for too many years. I had no issue with reinforcing a consultee with knowing the rights that their loved one felt he no longer had as a prisoner were not abolished and couldn't be infringed upon. The consultees' had the voice of their loved ones, especially the wives. So, use those rights to the benefit of your family. As long as they were not abusing the system and were advocating for entities to be held accountable for their wrongdoing, there was nothing that could stop them.

Prior to Hamza and I getting married, Proposition 57 was a law passed in the state of California. This law was definitely one that touted the importance of decreasing the number of inmates being housed in the prisons due to mass incarceration policies that had been established many years prior. Those policies were in effect keeping low level criminals locked behind bars, some with indefinite sentences. The Federal Courts began to impose sanctions against California State Prisons and there had to be something put in place to help California reduce its number of inmates. Prisons were so overcrowded inmates were sleeping in make-shift housing units such as gymnasiums or sent out of state in the hopes that the loophole would have been considered as lowering the numbers of inmates if they were located in neighboring states. But, it was found out that even outside of the state they were still being rightfully counted as a part of California State Prison population. That quick fix ultimately did the state no good and the federal government would continue to impose fines and sanctions against states that remained out of compliance with the law.

Proposition 57 also made a change that allowed inmates who, after evaluation at the counselor level, were eligible to be able to participate in the Family Overnight Visiting programs. This privilege had been barred in previous years from "lifers" in the 1990s due to an inmate's decision to cause bodily harm while participating in the program. Due to the breach of safety to the visitor and the likelihood of a "lifer" receiving any consequential punishment for an act of harm because they were already sentenced to life, it was easier for the state to determine that "lifers" could no longer participate in the program. Now, based on the climate change of the state of California and the reformative policies that were being established and some re-established, more merit-based programs were being reintroduced into the California State Prison System. They were incentives that encouraged inmates to rethink the choices they made throughout their period of incarceration. Being able to have visits with family members and re-establish bonds that were once severed by a system that had no care or concern for the families that were forced into incarceration right along with the inmate despite their non-involvement in any crime, but indirectly serving a punishment as well, were now able to repair some broken relationships.

This was how Hamza was thinking as well. He had been on a new path of staying out of trouble for four to five years before the new law and was already making strides toward staying write-up free and rehabilitating himself, not based on the prison's implementation of new laws, but a desire to better himself. So with this already established mindset and his finding out about the criteria to receive the family visits, he initially applied to visit with his children in an attempt to start gaining their trust and show them he was on the road they needed him to be on as their father.

Because I was not married to him at the time, I was ineligible to be considered for a visit with him, so I was not a part of his application process. He had waited out the timeframe for his counselor's response and he was optimistic that all was well. He did not receive a rushed denial as some inmates would receive because they had not completed any of the prerequisites to become eligible.

Hamza was slowly let down as he finally received a denial from the second level review in the Counseling Department. He was confused as to why he was being denied because he had none of the immediate denial infractions and no sexually based crimes against family members or children. He also had no write-up or 115 within the specified number of years. Plus, he had participated and completed all the milestone rehabilitative classes that were set by the prison to establish that he was actively working at being a model inmate. No one person was perfect, but he was actively trying to do what the system had set up for them to do to demonstrate that they deserved an opportunity to prove they were capable of the next level of responsibilities. This system had a definite way of reminding inmates that no matter what strides they made they were still just an inmate. Hamza was not an inmate that would allow the system to run over him, so he did as he knew to do and that was to file the proper documentation to appeal the decision; to have it re-evaluated by a superior of the person making the decision to make sure it was properly evaluated; and that there was no oversight on the part of the person that made the decision.

He filed the documentation, and unlike any procedure I had ever witnessed before the same person that made the decision, the CCII, Corrections Counselor II, was the same person that would re-evaluate their own decision. I don't know where this practice was taught, but I felt there should be a fresh set of eyes looking at the documentation when there was a request for review and not sent back to the same person's desk who initially reviewed it. Of course, they were not going to admit they were wrong the first time. Such a process of fair re-evaluation would take a large amount of integrity on a reviewer's part and unfortunately, that was a rare quality in the system, especially if the acknowledgement of an error was being initiated by someone looked at as being uneducated, lacking the ability to know their rights, was classless and less than humane. You know what they call "those" people—an inmate.

Hamza worked diligently trying to have his application reviewed by someone other than the CCII, but to no avail. No matter how many times he appealed it went right back to the same person and the stance remained the same. If the CCII said no once then no would be said a thousand times over. Eventually, I received the entire package in the mail and looked it over from the initial application and through the levels of denials. Hamza made efforts to meet with his CCI and get a better understanding of why he was being denied, especially since in the documentation the CCI had originally agreed and approved him for the visits because he met all the criteria. But, after he was denied by the CCII, the CCI's position quickly changed to side with his superior. It was an expected, but spineless position taken by an employee that feared retaliation if he took a stance against his superior even if he saw what they were doing was unethical, wrong, or malicious with intent to cause harm.

In the Army before I headed off to Basic Training I was taught the 7 Army Values: L-Loyalty, D-Duty, R-Respect, S-Selfless Service, H-Honor, I-Integrity and P-Personal Courage = LDRSHIP. I felt these and similar values should be taught throughout government entities because it was being highly disregarded in this system.

The main reason why the "good ol' boys club" were established and able to exist in prison was because speaking against wrongdoing was frowned upon and alliances were established with the green (guards) versus blue (inmates) ideals. And no matter what, right or wrong, you stand with the color of your clothing which mirrors the same mindset of gang culture. That was why when my contract with the prison system came to an end, I was happily able to walk away from my position, grateful that I didn't lose my dignity in the process. I was glad I didn't become blinded by money and overtime checks or sold my soul to a system that when they wanted to prance your butt out, five to seven guards would surround you and walk you right out their gates in the most humiliating fashion. There was no way I could stay in an establishment that didn't support doing what was right over which alliance you must stand with.

As I reviewed the paperwork I saw errors on both sides of the table, both from the counselors as well as from Hamza. I spoke with him about the actions he needed to take for his request to be reviewed again. I informed him of the specific errors made on the part of the staff of both the CCI and the CCII. He took my findings and made the necessary changes on his part. I even composed a letter to the Warden, alerting him to the lack of use of the chain of command in the appeals process. It made no logical sense why no one above the CCII was even being notified or involved in the situation. Needless to say, acknowledgment of an error is not something this facility took kindly to so I received a letter in response that lacked any responsibility for the error. Instead, they poorly plagiarized my letter to them and then requested that I speak to the inmate that they had mistakenly referred to as my son which showed me how little effort they took in paying attention to what they put in writing. Less than surprising, a system where, "the apple doesn't fall far from the tree." A tree that bears its own fruit.

After Hamza and I were married he made another attempt to apply for family visits and I was officially able to be included in the application, unlike the previous application. Unfortunately, the same two people who received the first application would be the same two to review this application as well. Some things hoped for would not change. This time Hamza was a bit more grounded in not getting his hopes up high, but he wanted to remain optimistic that he would be approved this time around, since his merits of behavior and positive programming participation awarded him this privilege. The new law loosely paraphrased stated that old infractions could not prevent inmates from family visits as long as they were not automatic disqualifiers identified in Title 15 California Prison's Regulation Manual. Going by the new law guidelines Hamza fit the criteria to be eligible. So, it was fair to assume he should be granted family visits, but based on the chips that were permanently lodged on the shoulders of the staff he was required to deal with their battle of the egos weighed heavier than a professional review.

I had the privilege of seeing altered documentation with false information written in the margins on official documents which had a clear outline to follow. I had witnessed multiple changes of what the reasons were for denial and every time the next accusation was refuted with evidence, a new accusation was presented. The one thing I could say was the most positive outcome of all this nonsense was the support I was giving my husband as he maintained his cool throughout this mess Hamza had even sat through a meeting with his counselor, the CCII, a supposed professional female as she yelled, "I don't want to see that document even if it was signed by Lady Gaga!"

If that didn't scream unprofessional, I don't know what did. We all knew if he had yelled back at her he would have been slammed to the ground, handcuffed and carted off to the hole where he would be awaiting a hearing and then eventually shipped out to a new higher-level facility with the label, "Staff Trouble Maker" or even worse, "Violent with Officers" as a forewarning to give him the welcome he deserved wherever he landed. However, her screaming only caused an Officer to peer around the corner to perform a quick check and then returned to standing guard as if nothing was out of the norm.

Hamza was truly learning to build his defense case, and I was fighting right alongside him. Far more official agencies had been placed on notice as to what was taking place than even CSP-Solano was aware of and they were on standby to see the outcome of whether or not this facility would uphold the law or as my grandmother used to say, "Cut off their nose to spite their face." The screaming woman was far more interested in seeing the documentation from the Inspector General that Hamza had in his possession, but she did not like his spontaneous reply of, "You shouldn't be interested in it since it wasn't signed by Lady Gaga!" They had already violated several of his civil liberties in this process and I told him to keep on letting them continue to pour gasoline on the fire. One thing was for sure, when they finished making this simple situation larger than it had to be, they would have learned many preventable lessons along the way. The person they believed to be uneducated, lacked basic rights and was just an inmate that had no support, would have proven to them all that being a human meant so much more than they gave credence to and it would be to their detriment. The letter the woman wanted to see from the Inspector General was in fact addressed to me. I was his wife and I had every last one of my rights and had no problem with using them when I knew there were violations within their ranks.

A common misconception and huge oversight on behalf of the average prison wife was their lack of having a Power of Attorney, a Living Will and Trust Accounts. Yes, they were married and that gave them a lot of control, but their husbands had not legally given them full access over their affairs in his absence without the proper documentation. I even noticed that wives lived seemingly separate lives from their husbands. There were no established bank accounts; business affairs were separate; and legal documents were not signed by both parties. It led me to question the stability of marriages. Is this happening due to the husband and wife's fears of committing and submitting to one another completely? What established connections do they have as a married couple outside of matching rings and an official marriage license? Those were common questions that arose for me. I often noticed in prison marriages there was this undertone and disconnect. Those were all things discussed openly between Hamza and I and the basic outcome of our discussions was a win-win situation—we win together and we lose together, so everything we did was together.

I entered the visiting room often with documentation because we had business to take care of. The old folks used to say, "All romance and no finance soon will become a nuisance." There were so many measures of protection available in this day and age. Even if a couple were afraid of the potential for divorce and their spouse taking them for all they were worth, then as in any marriage, then the couple needed to discuss the options of a prenuptial agreement. If they were not able to establish that form of security before the wedding day, postpone the wedding until both of them were more confident in the decision to move forward. I had told many friends, whether a prison wife or not, or any bride for that matter, "There is no race or deadline to the altar. If you are not ready, don't let anyone, including society, tell you there's a set time." Marriage was not to be taken lightly and in all things there were rules of engagement. Set your standards and don't waver from them. Never become a victim to your own ignorance or to being naïve. No love would ever be greater for yourself than that of your own.

CHAPTER TWENTY
Public Displays of Affection

We were taught that communication and trust were key components of a loving, long lasting, and healthy marriage. Well, a prison marriage is one that should be built of fortified concrete because those two qualities would be used more often than any other. Unlike a traditional marriage it possesses the luxury of constant connection through the senses and daily interaction so words can sometimes be left unspoken. In a prison marriage I found visual connection was the least utilized mode of interaction. Also, as learned in theory, if one sense was deficient there was an increased use of the other senses. A prime example was in a blind man, where he lacks sight, his hearing, smell, taste and touch were amplified to help him navigate through life. Prison marriages relied heavily on the need for effective communication skills which collectively would build trust. Knowledge of technology was happening fast throughout the prisons with the accessibility of video calls, but that has not been widely distributed throughout every prison, so there is still a lack in the sense of visual connection.

In a traditional marriage a wife could get up in the morning, put on her favorite blue dress and get ready to head off to work. Her husband would watch her strut through the house confidently because she knew she looked radiant and the dress would probably grab his attention. It might even send a signal to her husband that after work she wanted to do more than just cuddle and fall asleep. Well, a prison marriage was significantly different. Instead of having the convenience and luxury of my husband seeing me in my favorite blue dress, I would need to describe the dress to him and he might even have to refer back to some previously sent photos to hopefully come across the blue dress being referred to. This conversation would likely take place long after I strut through the house on my way out the door to work. I would likely receive some compliments from people whose compliments would never matter as much as the ones I was longing for from my husband.

And that opportunity for more than cuddling would be reviewed and monitored for safety reasons by Officers who indirectly participate in the intimacy of a married man and woman. They attentively listen in on whatever version of phone sex we liked to engage in. Does that stop them?

Hell no! Their intrusion into our intimate time was now a consolation prize passed down with the determination of his sentencing. These are the same Officers who will see my face and remember my name as I continue with my routine of visiting my husband. Now, those Officers were paying me closer attention than ever before because they knew more about me and my husband than I ever cared for them to know. Would this prevent us from doing what we wanted to do? It would not. With the prohibition of private conversations due to the enforced, no cell phone policy and our having to adhere to the rules in order for him to remain eligible to come home when that time comes, their listening in on our sexual talks was the least of either of our concerns.

We both believed that participating in the Family Visit program would make the sexual tension between the two of us lessen. But until then, we did our best to follow the rules in the visiting room with one kiss at the beginning and one at the end of each visit. We decided to just make them some of the best kisses imaginable. Hey, we gotta do what we gotta do. We also follow the holding of hands rule. I had to swat his hands away often enough and remind him that we could not afford the consequences of any perception of actions outside the rule. This was rough for two grown-ass adults who already knew what each other was like in bed. So, I was not guessing or assuming when I said, "Damn, I know what is in store when we are given the chance to be up close and personal." Hamza already knew the thoughts and desires dancing around in my head because I had not been shy when telling him. I check him out often. My mind and eyes would tell on me often enough, just as his did. Even with loose fitted clothing and a hijab covering up many of my private areas, he could see straight past them. His imagination and memory easily manipulated the cloth that covered my body. One thing was for sure and two things were for certain, our chemistry had never been in question.

There were some couples that had gotten caught when they couldn't contain themselves and I assumed the Family Visiting Program repeated denials had gotten the best of them as they tried to sneak and get in an extra kiss or touch. Some had gone further than that. Hamza often told me, "You see everything." I told him that situational awareness had been ingrained in me in the military so there was not much I missed. I would notice the attire of a visitor or the anxiousness and body language of an inmate or a movement and from that alone I could tell that it was obvious the couple couldn't hold out any longer. In the visiting room I saw subtle sex acts as well as the blatant ones, too. I don't flinch a bit at a couple who remained discreet. If anything I tried to give them their privacy and paid them no attention at all. The ones that bothered me were the ones that on occasion would lose all control and go for the gusto, right there for all to see. The act might have started off innocently, but they would lose focus about where they were and everything became exposed. Basically all of us were antsy and horny, but we hadn't allowed our desires to get to that point.

However, one fact still remained the same, it was a public room and kids could run around all over the place and I believed in safeguarding their innocence. So it pissed me off if I noticed someone exposed to the point where children could potentially see them. But usually when it got to that point, the Officers were aware, too, and would break up the scene. The latex gloves would come out and they moved across the floor quickly, zooming in on the location. The inmate would usually automatically rise, sometimes adjusting himself then head off to the back. The female sat and waited for her marching orders and escorted out through the grill-gate where she would be informed that her visiting privileges would be terminated. The cancellation could extend for lengthy periods of time, depending on what they caught on those cameras. The inmate that got caught in the act had some serious charges to face, especially if they determined he had exposed himself in front of children. So I always thought to myself whenever I saw a bust of that magnitude, *"You gotta be careful."* They walked a fine line between a fleeting moment of pleasure and getting hit with an "indecent exposure while minors were present" charge. That moment of intimacy was surely not worth it for the inmate or the woman.

I have come to find out that some Officers truly act as if they are envious of the relationships that exist in the visiting room. Their bias would show in their actions. Once they picked a couple to focus on, that couple would not find any relief from scrutiny until that Officer was no longer working in the visiting room. It was unfortunate to witness and I concluded that the Officer probably was not receiving that type of love and attention at home. I mean, what else could it be that caused them to dedicate that much attention towards a specific couple, especially one that was just enjoying their visit? Realistically, visiting was only two to three days out of a week at a maximum of six hours per day and they were exerting that much intrusion into trying to ruin the short amount of time that the couple was allowed to spend together. I assumed if you were happy at home you would be far less inclined to come to work and give someone a hard time.

A routine we saw and one Hamza and I laughed over far too often, we'd see the uptight Officer looking for an opportunity to leave the podium and walk over to engage with a couple. They are normally in an in-depth conversation and the Officer would come up with something as simple as the female's legs were crossed at the ankles and under the chair of the inmate as a reason to go over to them. The Officer would be approaching them as though they had made a significant drug bust just to tell the visitor to move her feet so they remained on her side of the floor. It truly got petty sometimes. It was bad enough everyone was subjected to sitting at those kid-sized chairs at preschool-sized classroom tables that were brushing up against everyone's knees. Then the moment a person found a comfortable position Officer Miserable-at-Home would come to save the day. That same type of Officer would find the most miniscule things to write-up inmates and terminate their visits for. I often felt bad for the inmate because no matter how bogus the write-up was, they would have a fight on their hands, trying to get the write-up removed. Proper procedures were not in place to give an inmate much room to argue against an unjustified accusation formed against him.

There are some Officers who are too laid back and busy trying to be accepted by everyone and that, too, is unacceptable. The inmates knew exactly who the Officer that overlooked everything was and that was the day all the transactions were planned to take place. I had my share of visits where I had to say to Hamza, "I hope we can get through this visit without this room being raided." Days like those made our visits really hard to enjoy because our antennas were raised and focused on what could happen. We did our best to ignore what was going on and stayed attuned only to us. We found a way to make every visit all about us, our family and the plans we needed to make to get Hamza home so he no longer had to live under those restrictions.

We were in full agreement that we were tired of our kisses being counted, the placement of our hands being monitored or someone pointing out that a hug seemed to be too excessive or too affectionate. Enough is truly enough with us not being able to operate like the adults we have become. We had experienced our teenage through young adult years of being able to dictate our own movements and how we interacted with one another to become husband and wife. So having someone dictating our actions between each other could quickly become frustrating and exhausting. I think I was over it after about two visits in. Yes, I enjoyed that incredible, mind-blowing first kiss in October 2015, but now my greed had increased. I wanted him when I wanted him and on my terms, not on the terms set by the Correctional Officers. That tall, dark and handsome man that I was married to deserved to be able to have me in any way that he wanted, but we did our best to contain the energy that was magnified whenever we were in the presence of one another.

We found creative ways to release our sexual build up. Aside from solely relying on calls which was most common to other prison relationships, we wrote detailed letters and intimate cards that I was sure when read by the Officers in the mailroom the descriptions would give their imaginations a full visual of sex scenes. Pictures also worked well. Prior to us being married I sent over 100 pictures that I had fun taking. I was thankful to his sister who didn't mind performing the favor for us. Her photographic ability reached the level of expert on that evening I took those pictures. I rented an upscale hotel room in San Francisco that was near the Bay Bridge. The lights of the bridge were reflected in the background from the balcony which made the pictures even better than I had expected. There were multiple wardrobe changes, positions and angles galore to keep him enticed for plenty of weeks. The impromptu honeymoon boudoir session had nothing on his engagement pictures. I got turned on when I looked back at those photos. That session would forever be remembered as, "Naughty Girl."

Of course, the ability to engage physically was limited and it became frustrating from time to time, but Hamza and I had committed ourselves to the long haul and established ground rules. If either of us was going through those hormonal moments where we needed each other's undivided attention, we would give it. We had not established any rules of stepping outside of our marriage to accommodate our sexual needs. There were no hall passes of any sort being distributed for either one of us. Some marriages had this allowance because the inmate felt it was selfish to deny their partner that access to relieve the tension. However, for us that was something he and I were highly capable of satisfying. I had plenty of sex toys to satisfy myself for any occasion or desire that could arise. In my past relationship with my ol' high school sweetheart who was in the military, we granted each other permission to have sex when we were absent from one another. But in my marriage to Hamza, that agreement was completely off the table. I respected his wishes and he respected mine, so there would be no outside calls to other women to satisfy something for him that I am fully capable of fulfilling. Our level of respect and level of communication with one another had boundaries that far surpassed the understanding of people outside of our situation. We understood each other and this rule did not need to be explained. Our relationship had enough pressure on it with the restrictions of incarceration so we did not accept any outside opinions of people coaching from the sideline, especially under circumstances that they might have never faced. The sanctity of our marriage was dependent on the two of us and our relationship with God and those were the guidelines we had decided to live and love under.

CHAPTER TWENTY-ONE

Prison Wives Club

Being a prison wife or a visitor in any capacity when coming to see someone incarcerated, there was a lot to know and many ways of navigating through the wealth of information that would need to be absorbed. Hopefully, there was someone you could trust to give you the ends and outs to make it through without many of the hassles. In the beginning the basics I got from Hamza was about what to wear and what not to wear to get me through processing. That was 17 years ago and we've since found out even that information changed from facility to facility. He had instructed me to rely on Title 15 and the DOM when I needed to question the legality of anything and to never rely on someone else's opinion. I took his advice and ran with that for many years and it played out well and often to my advantage.

The rules I had gotten very early on from Hamza had helped me tremendously throughout this journey and I knew they would last me until we were no longer in this situation. Rules such as, "Do not let anyone ride in your car unless I say it is okay" and "Do not make any friends that I have not approved." The average person on the outside looking in would look at those directives and immediately say, "Oh, he is controlling you! Girl, you don't need to do everything he says. You have a mind of your own and you can make decisions without him." But this was 100 percent not the case. There are established rules that were set and I respect them because they were set to both safeguard him and myself. Every situation you encountered in a prison was not the reflection of your situation and you had to be mindful of that. As a prison wife, during visitations, you didn't know what the couple sitting just seconds away from you was up to. Actions they might consider as cool in their relationship, you might not consider to be cool in yours.

In my previous marriage, my ex-husband was less protective of me and would ask me to bring people to visit. Now, just because the request was made never meant I was so in love I would follow through. And it surely didn't mean that I was not going to safeguard myself. Over time, I had seen many visitors' vehicles checked based upon someone offering a ride to someone. I saw women doing the "walk-of-shame," escorted by officers and placed in restraints with their children following behind dazed and confused, crying as they tried to absorb what was going on.

I also knew of women doing time because the district attorney built an air-tight case for the charges of distribution or trafficking narcotics into a facility with clear notifications that they would prosecute the offenders. What I was not about to do was get myself tied into all of that when there was a possibility that a person I am transporting drugs for might fold then I would be left doing the time on my own while fighting a case that would take forever to overturn. I can honestly say I haven't met a dick or a man with a tongue slick enough for that level of persuasion.

The fact remained; at the end of the day I was responsible for the actions that I allowed to take place. Therefore, I could not blame anyone but me if I ended up getting caught up in something I didn't want to be involved in. That was why there was also a charge of "accessory after the fact" that is still in the law books today. I was not about to place my freedom on the line for a relationship. Keeping it real ain't that damn strong for me! So I had no problem responding, "Hell no! Whoever it is will find a way there or walking however many miles to get there will give them plenty of time to figure out how stupid they are!"

Lots of people don't want to listen to old folks, but they had some great sayings that would forever be relevant:

"If you're going to rob a bank, do it by yourself then you only have to worry about you telling" and "Ain't no sex that damn good to be stupid over." I don't know how anyone else grew up, but I stayed under older people to gain some of their wisdom! And in my previous relationship he quickly became my ex-husband for similar reasons. If you are not willing to safeguard me, my children and what I stand for then you and I could be good friends without a license attached to us. I had never been that gullible or desperate for any relationship. Plenty of men have placed their bids in, so deciding to leave that marriage was not a difficult decision to make. I knew my worth and I placed value in that. Being in love is a great feeling, but take your head into it with you. Never deny your gut instincts of taking care of you, FIRST!

So, with the addition of technology and access to everyone the prison had caught up with the times as well and they had social groups and sites. Some wives, girlfriends and friends had established their own groups we all could join, interact and connect. There were some really good ones intended to provide resources and information to prison wives, families and friends. Shameless plug—I have my own as well! Mine was available only to provide services, I did not own or operate a community-group social site. I would check out resource and service sites and had no problem with utilizing them. They often provided plenty of researchable topics in one place for ease of locating important information. I appreciated the organizations that were offering services to assist in providing some normalcy to the crazy and unpredictable lifestyle prison families were subjected to as well. However, a social media site that was based upon building cliques I stayed away from. When I have tried to playfully discuss such a site with Hamza prior to being married, I had gotten back a very firm stance and a not-so-subtle hint of, "You must want to be with them and not me." I could read between the lines very easily because I understood the person I was in a relationship with and who he was. I knew neither of us wanted me to engage in something like that. But I just liked to get a rise out of him as I listened to his reaction because it was funny even though I knew he was serious as hell. It avoided the cattiness and conflicts that could ultimately affect him inside the prison so it was something we had agreed on for both him and myself long before we were ever married.

I made no judgment calls on groups such as those. I am not in them to even say what goes on. We had established in our marriage, time and time again, that our visits and our time were allocated for us and us alone unless it involved bringing our children, friends, or family. That was acceptable. However, making it a social hour in the visiting room was not going to happen, not without Hamza or I giving one another "the" look that needed no words at all. But we knew it meant, "You are doing way too much and you've got only a few seconds to fix it." We've had enough time apart so the time we did have for each other was simply for that. I was engaged in activities, jobs, children, family and friends during my free time when I wasn't visiting. Yet there were still times where I felt there were not enough hours in a day, even for a person who barely slept. So, my already busy schedule was an indication I didn't have any extra time to give to any additional outlets.

In those last couple of years, I had built friendships with a couple of women that stemmed directly from the prison. And yes, they were approved by Hamza or the relationships would not even exist. We were introduced through our spouses and I was permitted to speak to them outside of the pleasant smiles and sincere greetings or a simple hello inside the visiting room. We shared a common lifestyle as prison wives. We had no social media following or groups formed that we participated in. I engaged with them as I would with any close friend through text messages, phone calls, hanging out together during our available time; all the common things of any sister-friend relationship. The greatest part about those relationships was that we could relate and empathize with one another. We understood the language that we spoke without having to apply a backstory first. There was no judgment that we cast upon one another, and no disclaimers needed because we discussed the frustrations or positive changes we witnessed within the prison system. It was refreshing to have people that you could be open with. I had to trust they were not engaged in any of the other stuff that could take place inside the prison that would jeopardize our friendships.

They were similar to me. We were all in the same situation as we prayerfully asked God to grant us relief as we provided sound support to our husbands as we saw fit. I gave a big sigh of relief that we had each other. And for that reason I didn't knock building friendships with other women that were in a similar situation. But, as with anything, always be careful of the deception and manipulation that could take place because the consequences didn't just involve the two of you, it could become a problem behind that visiting room door as well. Gossiping with the girls took on a different look and effect when you were a prison wife. So I had learned over time to mind my own business and stay in my lane. That quick comment or careless thought you decided to share could ultimately turn into a brawl on the yard later on that same day. You never knew how things could escalate from something harmless to a big issue when it was misinterpreted by someone else.

The husbands have brothers they trust and some guys they just keep near, whereas we wives learned quickly the difference between the two. This was why there was that established agreement between Hamza and I that I follow his lead because I did not know what had transpired throughout the week in my absence, just as he might not know the same for me. Oftentimes, we were each other's eye and ears. There could be information he might not know as well; things I did not pick up on that he needed to make me aware of to insure we were both okay. As a team, we compared notes with each other regularly to make sure he and I were on the same page. We would hate for one of us to be aware of something and the other didn't know. We had faced this scenario in the past and had to quickly resolve the situation to make sure that our best interests were taken into consideration.

Effective communication and trust would always be keys to our marriage. They were skills that we were constantly working on, building up and fine-tuning as we grew. Hamza would sometimes hit me with a question such as, "Give me two things we are doing well and two things we need to work on." I could tell this was something he was learning in the counseling groups that he participated in. I appreciated it as a wife that he was attentive to us and knew that those were areas that we as a couple should remain focused on. There was nothing wrong with applying a few of the counseling and therapy strategies into our marriage to make sure we stayed on track. I had already expressed to him during our pre-marital dating phase, "What you did to get me, I need you to remain consistent in order to keep me." Many people would attribute a statement like this to dating rituals such as buying flowers, candy and cards for special occasions or special times for him to show he was the same man, but that was not what I was referring to. I didn't need Hamza to remain the same just as I'm not the same woman as I was in 2015, or the teenaged girl he originally met back in 1992.

People continuously evolve and I needed Hamza to be attentive to that so that we didn't grow apart, and I would do the same for him in return. I believed in the analogy that relationships were similar to cars. In order for a car to stay on the road you must perform continued maintenance. Do not expect a car to continue to run just because you filled the tank with gas and put air into the tires. It required far more attention than that. As the car aged the maintenance requirements would become more extensive because parts would start to wear out at a faster rate and you had to stay on top of things to at least see a repair coming and fix it before it was too late.

The last thing that should not be done in any relationship, including my own, was share what problems existed in your marriage with every listening ear, or everyone on the internet world. That was not going to go over well once Hamza and I had worked through our issues. Families could still be fighting mad while we had already decided to forgive one another and move on. If we did choose to discuss any issues in our marriage, we had a designated individual that we chose to share with and were confident that they would guide us in the right direction while also keeping private the sanctity of our marriage. It was the same person that performed our Islamic ceremony on our wedding day. We both trust that he would guide us back to what we had promised one another in our wedding vows. That was the most important and sound way of handling any disagreement that we couldn't fix between the two of us. Faith was what we relied on to get married and it would be the source when we couldn't see our way through. Sometimes a listening ear could be attached to a running mouth and not a source to provide reliable advice. So, if I had an issue within my marriage I would take it to a source I could trust.

My husband was not someone I would ever put down and ridicule in front of others. I hated when I heard about or saw people tear their relationship down in front of others. You could be mad all day long, but never defile your relationship like that in front of other people. I know every relationship could go through good and bad times, but tearing it up in front of the masses surely was not going to make it any better. And, you better believe what you've torn down in front of people, someone else was waiting to pick it up and step in where you were falling short. What's that saying, "One man's trash is another man's treasure?"

Hamza and I had some standard guidelines long before marriage that has helped us to this day. When it was time for us to do our homework prior to the wedding we were doing it. It was not just about making a beautiful wedding day. We wanted to focus on building a healthy, lasting and loving marriage one day. He had been married twice before and I had been married once before, so we had experience to pull our data from and made sure we were focusing on the areas that had high fail rates. At our age, we both had determined neither one of us was looking to get married again so we were going to work with what we had already invested in ourselves. And as I playfully, but seriously, liked to remind him when he was driving me insane, "I know the man I got is a keeper and I will fight you to keep him." So he knew I was not leaving this marriage without giving it a full, focused effort to succeed. As we have stated time and time again, "The only people that can damage the marriage we have is him or me not anyone else." With his big ol' strong arms and Chester Cheetah size grin plastered across his face, he pulled me in for another one of those confirmation kisses that let me know he's got me, and I got him!

CHAPTER TWENTY-TWO

Freedom is not Always Free

We had come a long way over the years we have known each other. We started as teenagers back on the block of my old neighborhood, immature and doing what we wanted to do and now, after a year, being married to a person I still refer to as my best friend. He had grown from a young, wild and crazy "Boo" Vandrick to a vastly different Hamza that I proudly married. He was a man who could say that despite the circumstances and the unfair justice system that threw the book at him to expeditiously close a case, he took the time to hold himself accountable and admit that he wouldn't be in this place had he made better life choices, not trusted people he thought would always have his back, and been committed and loyal to the streets that proved yet again that in the end, "They don't love you!" Also, what he had come to terms with was the life he was living was going to eventually dish him the same reality that many young men had succumbed to: "Live by the code you'll die by the code."

He lost his mother which was a devastating blow and not being able to see her body laid to rest took its own toll on him. He knew my promise to him was that as soon as he was released the first thing we would do was to take him to his mother's burial site. When you've served the amount of time that he had thus far, you had a different outlook on past decisions and life as a whole. So even though he hated the system, he also attributed the system to his survival and his growth. I could also say as his wife, I was proud of the man I got to know daily. He was truly not the same hard-headed, defiant person who passed through those large electrified gates laced with barbed-wire. He had spent the time needed to work on himself from the inside out and redefined who he wanted to become as a man and the man he was called to be.

God had taken us through some tests and trials that we hadn't spoken openly about to many others. Many had no idea the amount of stress and devastating humiliations inmates face. After a while many become numb to it. So as Hamza and I had discussed, we were not ashamed to admit that counseling would be an option in the future if we needed it to help us carry on during his transition back into society. I had walked the path alongside my husband when an inmate took his own life with a bed sheet right after visiting with his wife and daughter; and while another brother, just days later, went out on a medical code to never return as his last words spoken to my husband were, "I'll see you when I get back, Ahki." Things of that magnitude could never be removed from my husband's psyche no matter what anyone said about the grieving process. And as a committed fiancée at the time, I was able through the encouragement of Hamza, assist and make sure the brother who took his own life was properly buried as a Muslim. That family would forever be in my heart and prayers as we bonded over such a trying time in their lives. And the brother that went out on a medical code was the reason my husband was even in the Honors Program that he had been participating in for the past three years. That brother had recommended Hamza as suitable for the DeLancey Street Program, and his family would forever be in our prayers and hearts as well. He was a newly married man and for a long period of time it was difficult for me not to see him and his bride, smiling and laughing together, after his untimely death. It had been refreshing to see the glow on their faces as they greeted one another after a week of being apart. They would frolic around in the visiting room in their own little world, not at all phased by the external factors that incarceration could exhibit. Hamza had been one of the witnesses on his wedding day and I was glad he was able to be a part of that union. Before his death, they modeled the representation of what a happily incarcerated marriage could become for us.

Life threw us some unforeseen circumstances that we as a couple had to grapple with, dig deep to overcome and to see through; but together we had become stronger and wiser through this process. Those deaths taught us to never take life for granted and take items such as insurance policies, Living Wills and Power of Attorneys more seriously. So even before we were married we were making sure that we had things in order. As much as I knew what to do if something was to happen to Hamza he, too, knew what to do to handle my affairs. Those were things that the average couple likely had not discussed, but those types of conversations had become mandatory for us. The knowledge I provided to people came mainly from my many experiences of which most prison wives had not faced and some I prayed never did. However, if I could be of any help to prepare a person in the event they needed it, I felt as if my time was not wasted. I was willing to share that small amount of information and even the "how-to" of getting them started. My view was nothing is ever hard; as with anything it would take a desire and time set aside to see it completed.

It was not as if we had not faced our own scares as well. Just a few months back, California was facing fires burning from all directions. CSP-Solano and CMF Prison Vacaville were literally in the line of fire. I cannot forget that day, even if I tried. It was my 40th birthday and I was scheduled to meet with my mother for dinner. My daughter and I got on the road to travel to the other side of the Bay where we drove right into the thick of completely stopped traffic. I reset my navigation system to see if there was an alternate route we could take because traffic was going nowhere. I was redirected to take the back roads which would take me right past the front of the prison. A ritual that I performed any time I passed Hamza's prison was to blow a kiss in the direction of the entrance as I waved and called out, "See you soon!" or "Love you much!" The only difference about this day was that we were driving at a snail's pace because I was not the only one following this route. However, it was moving more than the freeway was so we crawled along through the grassy hills of a residential and industrial area.

My daughter was looking out the window at the line of never-ending cars and suddenly spotted a cloud of smoke in an odd location. I couldn't see it from my vantage point as she brought it to my attention. I replied, "Well, it's in this more rural area. Maybe someone is burning trash or the fire department may be doing a controlled burn." I thought nothing of it as the traffic was aggravating enough. We finally got a bit of relief and were almost reaching normal speed after about two hours of the slowest crawl ever. But, my daughter was receiving emergency alerts on her phone.

She said, "The traffic on the freeway is stopped due to an Officer involved shooting."

But, she was later notified that the fire I thought had been something simple was being reported as a grass fire and the neighbors in the area were being asked to evacuate, especially because the winds near the prison could get extremely high and without reactionary time. Now, I did notice that the hills where she first saw the smoke were fires burning right on the backside of the prison. I was hopeful those fires would not roll over to the other side and come close to the prison, but just in case I asked my daughter to set the same alerts up on my phone so I could get the alerts, too. She didn't hesitate. We continued on to dinner with the family and I was now being updated on my phone about the progress of the fire. The shooting incident was no longer being reported on as they got a handle on that incident earlier. It was getting late and I hadn't heard from Hamza, but I was sure the prison had an evacuation procedure in place if the fire got out of control. At least this was what I kept telling myself. I had been through a similar situation as an employee when I worked at the prison just a few months prior. When the Napa Valley area caught fire, we received inmates at our prison as a part of their evacuation procedure and even Hamza had told me they were preparing them to receive outside inmates into their prison as well. He said he was more than willing to share his cell if needed because he would hate to be in that situation himself. Now, here we are with his prison possibly under threat from the fires.

I received a call from my cousin while I was at dinner. Her husband was at the same prison as Hamza, but he was on the other side. She had heard from her husband but hearing from him had only made her more panicked. I started trying to reassure her that everything would be okay, even if I wasn't 100 percent sure of it myself, but I would never encourage panic even if things were not okay. I knew panicking would not help at all. I was able to calm her down enough for her to tell me, "My husband said they are locking them in their cells and they haven't been told anything about leaving or preparing them to leave." I didn't have any official information on what Solano's protocols were, but I did not believe they would just let an entire prison filled with inmates and Officers burn to death. At least where I worked that was not the information that we were given in the fire safety meetings we had. So I was able to let her know that.

I was able to get back to dinner, but that didn't last long because my phone rang again and it was one of my prison wife's friends and she started giving me an update she got from a friend of hers. Visiting had been cancelled for the next day. She was unaware of what the protocol was going to be, but information was being sent out to families telling them not to come near the prison because the fire was definitely close. I was still receiving notifications from the Emergency Alert System as well and the fire locations they were providing was definitely too close for comfort for me. I was praying the firemen could gain control and those winds would die down. Not only was I worried for my husband, but I also lived close enough that this was also a time for me to think about my residence as well. It was a small thought but I knew my homeowner insurance payment was current; there were some irreplaceable items that I would hate to lose. However, it was far less important than the fact that I was grateful that my daughter and I were not at home.

However, how was my husband? God, I couldn't help but think I prayed they were ready to evacuate. But I refused to break under pressure so I did my best to remain calm and let my friend know I was keeping track via the alerts and if I found out anything I promised to call and let her know and she promised me the same. Luckily, I did start to receive notification that the winds were easing up. The fire near the prison was shifting and the firemen had control over it. Cautiously, evacuations in the area were slowly being lifted, but any small amount of good news was a million times better than the possibility of bad news. I sent out the text messages that I promised to my cousin and my friend, hoping to bring them some relief as well. Both assured me they were following the news on TV as it was being reported there, so they were getting updated as well. The news was much better than before and after dinner I felt it was safe enough for my daughter and me to head back home.

The traffic was clear, but it was dark as ever and I couldn't see all the damage that had been caused. I could see that the hill near the freeway had been burned. That was surprising to me because I knew that was a long distance away from where my daughter initially saw the start of the fire. We went home and I knew I wouldn't see Hamza in the morning, but before bed I prayed that God would keep him safe from harm. The other thing I knew was he and our son had a similar medical issue and that was that they both had asthma when they were younger. Although they both grew out of it, anything could still trigger difficulty breathing for them. So it was an unspoken fear I had about the quality of the air for him during the fire.

The next day I awoke and made sure to notify his daughter that visiting was cancelled and I continued to keep up-to-date on any news regarding the prison. Today, the roads my daughter and I had traveled the day before were shut down completely so no one could get to the prison even if they tried. Later on, I was fortunate enough to hear from Hamza. I lit up like a light when I saw that number come across that phone. Oh my goodness! I always said I never knew how much stress I could bear until the storm was over. It was a bad habit I had. When I saw his number scroll across my phone it was like someone had lifted an 18-wheeler off of my chest. I could breathe again. I was elated to hear him say his name on the recording and the recording needed to shut up so I could hurry and accept his call.

We talked and he let me know how bad it had really gotten; how they had no idea if they would live to see the next day or what was to come. He told me the Correctional Officers did lock them in their cells with no instructions regarding letting them out to send them to another facility if things got worse. The facility shut down all non-essential power so the inmates lost the small amount of circulation they had that moved the smoke around in their cells from personal fans. He saw and heard the staff leaving, as they avoided any questions asked by the inmates. He said he looked out the window and could see the fire coming down the hill and it came so close that it was on the facility grounds and touching the back gates. The only thing he knew to do was complete his prayers, he took a t-shirt and placed it over his face and then he lay on his bunk on his back. He felt if it was his time to go, he had no choice in the matter if it was God's will. I definitely didn't want to hear that kind of report, but I had to accept his words. I let him know the situations that were taking place on the outside and what I was going through, but I didn't want to dwell too long on the negatives. I needed to focus on the positives and the major one was that we were on the phone and he was alive to talk about his experience. I wanted him to talk to his daughter so I bridged her on the line and she was able to talk to him. Although things worked out, it served as a definite reminder that his sentence could suddenly turn into a death sentence unlike anything the judge had handed down.

People in traditional marriages tended to wonder about the concerns that many prison wives had learned to push past or had thought about early on. However, those concerns no longer served a true purpose to stress over on a daily basis, such as, "What if he leaves you once he is released? I had told those who brought their concerns directly to me, "If that is God's will, please understand that I will be okay. That is not a desire of my heart, but one thing I knew for sure and two things were for certain, during the time that I had been married to Hamza I smiled more than I had with many others. I have had more good days than bad. So I would leave the marriage knowing that I regretted nothing. He was mine for the time I was allotted. My goal was forever, but that was not up to me because when you have faith all things are left up to God who has the final say and I trust the plan He has already prepared for me."

Hamza had already been more of a husband than I ever expected him to be as I, too, had experienced a loss that was devastating to me on a personal level. My old high school sweetheart, "the Military Man," as some referred to him, had passed away unexpectedly and it had a great effect on me, but Hamza stood right by my side, stronger than I expected him to be. He knew the love that I had for this man went all the way back to our teenage years and not once did he make me feel guilty about my feelings that I had for another man. He was not jealous in any way shape or form. Hamza not only supported me through the process, but for a man that had little to give he made sure that he pulled every string he could think of and told me, "I will not allow you to sit at home and grieve alone. I am going to pay for you to travel wherever his services are located so you can say farewell properly as both him and I know you need to." A man that stood that firm in who he was and supported me as he did, I would forever be grateful, because I knew some men couldn't respond as mine did. Their feelings, pride and male ego would not allow them to be that transparent for their wife's sake and her need for closure. The other thing he said to me which was endearing as well, "I wish I could be there to travel with you as your shoulder to lean on." And, during my trip he called consistently to check on me and my old sweetheart's family. Upon my return he still checked on me regularly to make sure I was doing okay. I was so grateful for the man God granted me. I couldn't have prayed for God to send me any better man than what He had given me.

I don't even recommend or encourage prison life for the average person because that life comes with challenges that many are not equipped to handle. Hamza was far more than my husband. He was genuinely my friend. I thank God for him every morning, every evening and anytime in between! I have no shame! In my title as a "prison wife," being incarcerated is only the place where my husband resides, temporarily, because he has a home with me and our children where we would honor and would accept him with open arms soon enough. *Inshallah!* No one had to understand our journey or had to be happy for us. We do a grand job of it by ourselves. People could make up their version of events and tell lies to make their own roles seem unflawed, but we all knew the truth. No one was perfect. We all had fallen short of that, but Hamza and I were perfect for each other. He was my best half and I am his good thang. As we both had already stated in our own way,

"The reason that the relationships of our past didn't work out with any others is because we are designed for each other." It took us years to figure it out and that was okay as well. At least we found out.

The laws were changing and we had motions currently sitting in the court. Only God knew what the outcome would be after everything Hamza, the team and I had worked on was considered by the court. However, we knew for sure that along with the prayers we were putting in the work. We had a desire to win! We knew that the case and trial were rigged and we had accepted that, but it was time for the system to acknowledge it and correct their errors; to go back and admit the failures of the system that contributed to this mess. Justice should be served because it was supposed to be served back in 2002 through 2004. No person should ever sit in a prison for a crime they did not commit. Prison reform was taking on a life of its own and as his wife I prayed that it would reach all the way down to not only my husband, but other inmates that were sitting in similar situations as well. There were criminals breaking laws every day so if the governing officials were concerned that for every empty bunk they would lose money. They must realize that a new body was coming through the gates to occupy that space on a consistent basis. The prison system might have a decrease in funds as it corrects its mistakes, but it surely would never go broke!

A LETTER TO MY LOVE

Peace and Blessings of God be upon you!

You took my hand in marriage and we both said I do! I do to a life filled with uncertainties, but a promise to stay committed no matter what. I take my vows as seriously as I take an additional breath every day. I know at some point we will be able to wake up and see each other's faces. That day is not now, but it is coming if God says the same. I am your Prison Wife for the time being, but our love is not incarcerated. We share the fullness of any other fruitful marriage and this is our test. It is becoming our testimony. We've survived the challenge from the start. Every marriage shares the vows: "For better or for worse until death do we part…"

Husband, this is our worst and we are passing the daily test with flying colors. We share in the good and bad days and have not written each other off. We communicate because we have found this to be the strength of our marriage; we have found the source of our union and that is in God up above. What more can we ask for and not appear greedy? Our love is abundant and everlasting. You remind me constantly of my attributes and what you admire most. We even spend hours repairing areas that need our undivided attention. Things may change slightly as you are released, but God will grant us the resilience to work through that new transition. We shall work through it with white gloves just as we have with all other things. Be gentle, be delicate, promise to be compassionate and consider the feelings and thoughts of one another. These characteristics have carried us this far and they will continue to keep us strong.

One thing is certain; I will get to lie comfortably in the arms of the man that I desire and rest my cares on you. We can awake rejuvenated and laugh as we always do at the simplest thought of *"How in the heck did we get here?"* Hamza and Zoya, or more officially, Vandrick and Qiana are husband and wife. No one would have ever speculated this would become a reality, including the two people most closely involved—yes, you and me, two teenagers just having fun who became parents under the craziest of circumstances. And now, years later, we have blossomed into the cutest, most loving and respectful couple than either of us could imagine. As we have been taught, "God has one heck of a sense of humor." Only He has the ability to make a way out of no way. So we are stepping out of His way and allowing Him to be God all by himself. I wanted to take this time to say, "Thank you, Hamza for all that you have been to me. Thank you for holding my hand and assuring me that it is safe with you. Bay-bae, we got a good thang going on!"

I have shared some of the intimate details of willingly enlisting into my own version of incarceration through the military and the journey of becoming a prison wife with people I never expected to, but our love is worth every word written. The parallel images of our individual lives unfolding into a romance unseen and softly spoken. It proves that even through concrete walls, barbed-wire and electrified gates, if there are two people willing to see past the exterior obstacles a genuine love can grow and mature like a rose that pushes through the crack in concrete.

I pray you find rest tonight and awake restored to see a new day, closer to your release than the day before. Know that I am here with you right by your side. ♥

Sweet Hugs and Kisses.
Forever and Always, Your Wife!
Qiana Riley-Jones

By Qiana Riley-Jones

FAREWELL MY FRIEND, UNTIL WE MEET AGAIN

To my old high school sweetheart, "Mr. Military Man"

I trust God has granted you entry into His kingdom to find relief. You were a great friend, a mentor and an honorable soldier. You served this country for 20 grueling years. Your repeat deployments into war torn countries would take a toll on any woman or man. So I thank you for every time you laced your boots, grabbed your gear and secured your weapon to face another day! The things you shared with me; I will cherish them always. At such a young age you became family to me in multiple ways. Even though things didn't work out as planned for us, we never lost that connection and remained consistent in maintaining our friendship which was worth fighting for.

You taught me a lot. I thank you for walking me through enlistment so I would learn to lace up boots just as you had done. I am grateful you allowed me to see this world from the same lens as you did. You gave me a great understanding and appreciation for a life well lived with no regrets. Your personality helped to teach me to take on a raw and uncut persona that was needed to secure my place in positions that may have seemed too large for someone such as me. You made sure that I exuded courage and confidence instead of fear no matter what I faced. I admit I was meek but you helped me shed that tendency. I will forever laugh at your frankness about everything. I thank you for introducing me to and welcoming me into your family.

My heart is definitely broken that you are no longer on this earth but, I truly understand it was time for you to part this life and transition on. Know that I listened attentively to every word you ever spoke to me and it made sense then and makes even better sense now. I will do as you last advised me to do: "Do anything that you want to do because you deserve to be your happiest." This world was unprepared for a man such as yourself and only heaven can absorb your abundance. You are missed in the present, but I know you are looking down upon each of your loved ones that remain in this life. You protected the United States Constitution in your choice of profession and you were relieved of that duty upon your retirement. Now you have gained the best seat in the house to continue on with your responsibilities of protecting and guiding your family and friends without any hindrances. I know you will carry on this mission with great honor— At Ease Soldier!

I Thank you Forever!
Your Friend,
Mrs. Qiana Riley-Jones

DEDICATION

This adventure was possible through the gifts of writing and expression bestowed upon me by God. With the allocated time requested from my family—my husband Vandrick “Hamza" Jones, my children Qivan Riley and Jasmin Clewis—and their encouragement and support for me to continue and seize the moment, helped me to believe that I had it in me to produce a great outcome. It was just what I needed to carry me through to the finish line.

Gratefully, both of my parents, Latonya Lockhart and Glenn Riley Sr., have been able to embark on this new and fulfilling experience with me. I am hopeful they are pleased in witnessing me continue to grow and explore new realms and adventures in this lifespan.

Also, I’d like to give a personal thank you to my “Sunny Gazes,” Sungayaia Harris, for believing in me. As a true sister-friend you encouraged and pushed me past my fears, even when this was just a thought process. With your constant visions of what success and completion looked like you motivated me, and for this I am eternally grateful.

I am thankful for publishing my beliefs, experiences and insights and that they will become a valuable resource worth sharing with many readers.

A huge thank you and countless blessings to the editor, Full House Typing & Editing, for the hours you dedicated to make this manuscript a full and complete reality. I am grateful for your expertise, professionalism, and attention to details. Your work and our continued communication ignited a trust that my words were safe in your care.

Readers and supporters of my works, thank you all for spending time with this published piece. A sentimental novel that shares many layers of the journey of my own personal experiences. It taps into many levels of personal growth while remaining committed to the maintenance of a fulfilling marriage to someone that is confined in prison. I truly hope you enjoyed this work of expression.

With all Sincerity,

Mrs. Qiana Riley-Jones

ABOUT THE AUTHOR

~ Mrs. Qiana Riley-Jones is a wife, mother, military veteran, author and businesswoman ~

Wife: To my very charismatic, courageous, supportive and loving husband, Mr. Vandrick "Hamza" Jones.

Mother: Of a blended family consisting of two biological children and five children by marriage.

Military Veteran: A former US ARMY Sergeant who proudly served as a 68C (68WM6) and a Licensed Practical Nurse/Licensed Vocational Nurse with an Honorable Discharge.

Business Woman: I am the business owner of Qiana Compositions, LLC. I have a love for writing and have designed a career in line with my passion. I also achieved my Associates Degree in Liberal Arts/Behavioral and Social Sciences with an emphasis in nursing. And, I am a middle-aged professional which is pure proof that you can achieve anything you set your heart and mind to do as well as wear multiple-hats in a range of fields. Never be afraid to tap into your talents and don't allow anyone to speak about what you cannot do. Be the author of your own life, dictator of your own path. It is worth the ride!

Author: I am currently releasing *Broken Pieces*, a self-help/self-realization book inspired by my many friends, family members and myself who have suffered through some very traumatizing experiences and work daily to mend the wounds of our past to build a healthy future. Learning to not allow the past to define us, but shape us through healthy thoughts, concepts and all the available resources. Available Summer 2021

Creator of: No Shame! Prison Wife. Some Choose This Life, Some Inherit It Journal 1 & 2 Ed.

Consultant: No Shame! Prison Wife Community. Helping families and friends to adjust to the abnormal and normal aspects of living through the experiences of surviving the sentence of a loved one's incarceration.

Made in the USA
Las Vegas, NV
30 October 2026

56882186R00246